MICHAEL QUETTING

Foreword by **STACEY O'BRIEN**

PAPA GOOSE

ONE YEAR, SEVEN GOSLINGS, AND THE FLIGHT OF MY LIFE

Translated by **JANE BILLINGHURST**

GREYSTONE BOOKS
Vancouver/Berkeley

First published in North America in 2018 by Greystone Books
Originally published in Germany as *Plötzlich Gänsevater*

The translation of this work was supported by a grant from the Goethe-Institut which is funded by the German Ministry of Foreign Affairs.

18 19 20 21 22 5 4 3 2 1

Greystone Books Ltd.
greystonebooks.com

Cataloguing data available from Library and Archives Canada
ISBN 978-1-77164-361-0 (cloth)
ISBN 978-1-77164-362-7 (epub)

Copyediting by Shiarose Wilensky
Jacket design by Nayeli Jimenez
Proofreading by Alison Strobel
Text design by Shed Simas / Onça Design
Cover photograph by colourFIELD tell-a-vision.
Photo Credits: Björn Klaassen 2 (bottom); colourFIELD tell-a-vision 1 (bottom), 2 (top), 3 (top), 6 (top), 7 (bottom), 8 (top), 12 (bottom), 14/15, 16; Michael Quetting 1 (top), 3 (bottom), 4, 5, 7 (top), 8 (bottom), 9, 11, 12 (top), 13 (top); Picture Alliance 10, 13 (bottom); Tobias Gerber: 6 (bottom).
Printed and bound in Canada on ancient-forest-friendly paper by Friesens

We gratefully acknowledge the support of the Canada Council for the Arts, the British Columbia Arts Council, the Province of British Columbia through the Book Publishing Tax Credit, and the Government of Canada for our publishing activities.

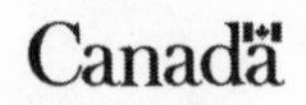

For Amélie and Ronin

Contents

Stacey O'Brien

Foreword

WHEN YOU LOOK UP TO SEE A *V* OF FLYING GEESE, DO YOU wonder if they have personalities? Or if they have their own separate opinions, feelings, and quirks?

Before I had the chance to raise a barn owl from a hatchling, I figured birds were all pretty much the same. I suspect that Michael was just as naïve when he accepted responsibility for an incubator full of goose eggs, even talking to them before they hatched, as their mothers would have done.

I could relate to his experience of raising, bonding with, and then having to let go of his brood. I laughed as he discovered he had to keep his hand in the box with the babies at night or they became hysterical, just as I had to do with Wesley.

I soon discovered that Wesley wasn't just "an owl," he was a sassy, opinionated, loving individual who happened to come in owl form. Michael, too, found that each of his seven geese had its own separate ways. Some were shy and stayed close, others craved adventure, and at least one tested him to his limits.

Even though he had to teach them some things, he found that they immediately knew how to do others—but he had to figure out which was which. As babies, they knew how to spread the imaginary oil from their undeveloped preen glands over their feathers to "waterproof" themselves. In the wild, however, a baby goose gets oiled up while sitting under the parent. Michael couldn't provide that to them, so he had to be careful not to let "his" babies to play in the water for too long.

Michael worried. What if he missed something important? How does one man teach them all they'll need to be able to live in the wild? The goal was to release them after testing small sensors that track how birds fly and how they react to weather. He had to train the geese to follow his ultralight while allowing them the wildness they needed to return to their own kind. It was a joyful but bumpy ride. They had to be coaxed down from roofs or tracked down in cornfields, but eventually they all flew together, close enough that Michael could touch them. I would have been thrilled to have the joy of flying with Wesley beside me, as Michael did alongside his charges. To soar with those you have nurtured since before they hatched must be exhilarating.

Once these little souls have captivated you with love, it's terrifying to let go. Because your friend is vulnerable, and you've shared everything, can you risk him taking to the sky alone, when there are so many dangers? What if he's shot? What if he can't make it through migration? Did I teach him everything he needs to know, you wonder? And—will he ever come back and visit?

When I accepted Wesley into my care, I had no idea how being with him would affect me. Michael went on a journey that few have experienced—but those who have are forever changed.

STACEY O'BRIEN, author of *Wesley the Owl*

1

Nine Eggs

I'M HEAVILY PREGNANT WITH NONUPLETS. AT LEAST THAT'S pretty much how I feel right now. My due date, the thirtieth day of incubation, is still more than a week away, but the nesting urge already has me firmly in its grip. I'm seized by the desire to break out into frenzied activity and research suppliers of premium-quality organic hay and grains. I would love to assemble something, prepare for the new arrivals as best I can, but unfortunately, I have absolutely no experience when it comes to building nests. The only thing I can do right now is to sit glued to the incubator.

Behind the protective glass lie nine goose eggs. The incubator is in the basement of the Max Planck Institute for Ornithology in Radolfzell, Germany, where I work. The contraption looks a bit like a convection oven, and it sounds a bit like one, too. The warm air inside is constantly circulated to distribute it evenly, and the machine makes a pleasant low-pitched humming sound. It may sound like a convection oven, but it's just 99.5 degrees Fahrenheit in there. No more, no less. A couple of degrees warmer and the eggs would be cooking instead of incubating; a couple of degrees cooler and they'd be in a state of suspended animation, as though they really were in a cooler.

The eggs are about as large as my fist. There are people who fry these distinctively aromatic eggs—one fills the whole pan—or routinely use them in baking as though they were hen's eggs packaged in bulk. That's probably a practical approach, as one goose egg is about the size of three hen's eggs, but I've yet to try it myself.

Inside the incubator, humidity fluctuates between about 65 and 70 percent. You could say that the whole thing is a fully automatic underside of a goose. The eggs are supposed to develop in the incubator exactly the way they would develop under the rump of mama goose in the wild. Unfortunately, this isn't as simple as it sounds. The underside of a goose is, in fact, an intricate anatomical wonder that combines moisture and warmth to create a precisely calibrated environment. If the eggs in the incubator are to hatch, a whole array of hatching parameters must remain constant all day, every day.

The most important of these parameters is humidity. If it's too low, the membrane inside the egg will dry out and become leathery, which would make hatching much more difficult for the baby geese. They might not be able to break through the membrane at all. Then they wouldn't be able to get out and would remain trapped inside their leathery eggs. I would prefer to spare my goslings such trauma.

In the wild, humidity remains high because mama goose leaves the nest once or twice a day to take a quick dip, and when she returns, she settles back down onto her eggs with a wet bottom. While she's away, the temperature in the nest also drops. Because the incubator is unable to replicate the mother's little excursions, it falls to me to take the eggs out twice a day, lower their temperature in cool air for half an hour, and mist them with lukewarm water before replacing them.

The eggs are creamy white and pleasing to the touch. They're warm and completely smooth, almost like a heated stone that fits comfortably into the palm of my hand. Every time I take an egg out of the incubator, I'm worried that I might drop it onto the concrete floor. By its third week, a smashed goose egg on the floor would certainly not look like something that could be cleaned up with a cheery pass from one of those kitchen cloths you see advertised on TV. It would look like an embryo, and the embryo might even be moving.

In addition to cooling and misting the eggs, I have to turn them multiple times a day. It's important to do this to prevent the tiny embryos from getting stuck to the shell wall. If they are to develop and grow, they must be free to swim around in their yolk sacs at all times.

The first few weeks, I was relatively relaxed when I monitored the parameters in the incubator, but now an increasing level of brooding paranoia is taking hold of me. When I'm at the Institute, I have to control myself to not jump up every few minutes to check the temperature and humidity levels. The night before last, I sat right up in bed a little after one in the morning and drove all the way over from home, because I was terrified that the eggs were suddenly too cool.

• • • • • • • • •

AT LOT IS RIDING ON THE PROJECT. AND BY THAT, I DON'T mean just the lives of nine little goslings. There is also the money involved and the success of my work. The goal is for the geese to eventually wear what are known as data loggers on their backs, matchbox-sized measuring devices that will capture a wide variety of information. The measurements will help researchers build a precise picture of the flight mechanics and aerodynamic adaptations of the geese, as well as capturing real-time atmospheric conditions.

If the experiment is successful, in a few years, or a few decades, it should be possible to use devices mounted on birds and other animals to gather meteorological data, such as wind speed and direction, from many different places around the world. This information would be automatically collected by satellite and relayed back to earth for analysis. These measurements would be highly valuable for weather forecasting, which usually depends on measurements taken at ground stations. For example, right now meteorologists can only speculate about and estimate what winds are doing ten thousand feet above Mongolia. One day, it might be possible to use birds as mobile weather stations to gather this data without affecting the way they fly.

If I don't succeed in hatching these eggs, that will mean a whole year lost, because there won't be any new eggs until next year at the earliest. Geese are not like chickens, which lay year round. Geese lay only between March and May, with the exact timing dependent on the weather. And perhaps that's one reason some people consider goose eggs to be delicacies: they're not available year round, and nature sees to it that the supply is limited.

It's the same for the geese, by the way. For example, if a hungry marten spots a clutch of eggs when mama goose is off for her swim, all she can do when she returns is mourn her loss, because she can't lay more eggs. There's no such thing as "double-clutching" for geese. I don't know if geese actually mourn, but a mother goose who has lost her eggs has no option but to wait a year.

• • • • • • • • •

BUT HOW COME I WAS CHOSEN TO BE FATHER GOOSE? IT'S quite simple: because I know how to fly. I've flown gliders for a long time, and a while ago I got my pilot's license to fly ultralights, as well. And that's why, when there was a discussion at the Institute about who should take this project on, it soon became clear that I would be the one bringing up the geese. And if all goes according to plan, in a few weeks I'll be flying with them.

The responsibility and suspense are weighing heavily on my shoulders. To say nothing of the fact that I'm newly divorced and still haven't really recovered from the stress of the breakup. I mustn't neglect the goose eggs, but I also don't want to take time away from my children, who are particularly in need of my attention right now. I just don't know if I have the emotional resources I'll need to look after nine demanding, peeping little baby geese who'll be hanging around me all the time.

For all these reasons, I'm edgy and out of sorts as I keep watch over the incubator. If I'm honest, one of the reasons I'm reading out loud to the eggs from *The Wonderful Adventures of Nils*, a famous Swedish story about a boy who flies away on the back of a farm goose to join a flock of wild geese, is to calm myself down a bit.

Yes, you read that correctly. I'm reading to the eggs. I place a Bluetooth speaker among the eggs and soon hear my somewhat-distorted voice.

"Once there was a boy. He was—let us say—something like fourteen years old, long and loose jointed and towheaded. He wasn't good for much..."

The important thing is that the goslings get used to my voice while they're still inside their eggs. Although they're not yet out in the world, they can still hear sounds.

When they hatch, they'll remember the sound of my voice. The sound of their mother's voice is also familiar to human babies before they're born. Apparently, there are particularly success-driven parents who press headphones to the pregnant mother's stomach and play classical music to their baby for hours on end in the hope of increasing their child's IQ.

"Don't worry," I say to the incubator. "I won't try to fly on your backs. I have my own plane."

The geese are definitely not going to get more intelligent listening to the sound of my voice. It's all about getting them to imprint on me.

Very generally, you can divide birds into two categories: precocial and altricial. Both hatch in the nest from eggs, but after that, their paths diverge. Whereas altricial baby birds spend some time in the nest after they hatch and are fed food regurgitated by their parents, precocial baby birds exit their eggs already well developed. Greylag geese belong to the latter category. As soon as they hatch, they're able to make their own way in the world. Even so, greylag goose parents still protect their young and stay with them for the first few weeks of their lives.

What's most surprising is that the baby geese aren't picky when it comes to choosing their parents. They accept as their mom or dad whom or whatever they're first aware of after hatching. Of course, in the normal course of events that's mama goose, because she's usually the one sitting on the eggs, watching over the nest after they hatch, and taking care of her little ones. However, the behavioral scientist Konrad Lorenz found out quite some time ago that geese will also accept members of another species as their parent—a human being like me, for example—or even an object, such as a soccer ball or a doll. All you need to do is make the baby geese aware of the object or person before and after they hatch. And that's why this process is called "imprinting."

Imprinting involves not only sound and physical appearance but also smell. For a while now, I've been putting a T-shirt I've worn into the incubator next to the eggs. An old sock might also work, but I don't want them to have to put up with *that* just before hatching. And I'm not only reading to the eggs about Nils; I'm also talking to them about myself, letting them know whatever's on my mind.

First and foremost, I'm imprinting two important sounds on the eggs, both of which are soon to loom large in the goslings' lives. The first is a recording of the propeller of the ultralight I'm going to fly to accompany the geese when they're airborne. They need to get used to the rattle of the propeller even though they're not yet hatched. The second sound—which is perhaps even more important—is the honk of a small somewhat old-fashioned metal horn with a black rubber bulb. The sound it makes is something like the honking of a vuvuzela, only not quite as loud but just about as annoying. Later on it's going to have just one meaning for the geese: “Watch out! On the double! Come over here right now!”

Wild geese make a slightly different but similar sound, which mama goose uses to warn her little ones of danger and call them to her. The world is a dangerous place for baby geese. Of the four to six eggs a mother goose incubates, often only one or two make it to adulthood.

“I’ll protect you from predators,” I promise the eggs.

They don’t respond.

I look at the temperature gauge again. Exactly 99.68 degrees Fahrenheit. Everything is as it should be. Then I look at the clock and get a fright. Hours have passed and it got dark a long time ago. Down here in the basement, I’m as cut off from the world as the geese are behind their shells. I suddenly realize how hungry I am. I jump up, bid the eggs good night, and turn off the light in the incubation room.

• • • • • • • • •

ONE DAY LATER, THE EGGS MOVE HOUSE FOR THE FIRST time. They stay in the incubator, but I move them from the turning rack to the hatching drawer. The turning rack is a device that automatically turns the eggs in the first few weeks of incubation. The contraption looks at bit like a tiny European-style clothes drying rack without legs and with a little less distance between the lines strung across it, which in this case are wooden rods. As the rods rotate automatically, the eggs turn gently on their own axes.

So soon before hatching, however, turning is not only unnecessary, it's also potentially harmful. For one thing, the goslings are now so big that there's hardly any space for them in their eggs, a tight fit that will be familiar to many pregnant women. For another, the goslings can no longer turn in their yolk sacs, because they've already pierced the tiny pocket of air found at the end of every egg. The air pocket provides the goslings with air to breathe before they break through their eggshells.

Turning the eggs at this stage serves no purpose and could damage the goslings. That's why I transfer the eggs to the hatching drawer, which is basically a box with a fine mesh screen that looks a bit like a lightweight perforated baking tray. I carefully lift each egg off the turning rack and tuck it into its new quarters. After their move, I no longer need to cool the eggs. Mama goose rarely leaves the nest toward the end of the incubation period and instead stays sitting nice and tight on her eggs.

It's important that the humidity in the incubator now rises slightly—up to at least 80 percent. The atmosphere under mama goose's downy nether regions is warm and moist. The membranes in the eggs must stay as soft as possible so that the goslings don't get stuck inside their shells or injure themselves while trying to get out.

Unfortunately, it's usually bone dry in the Institute basement. The whole time the eggs have been incubating,

I've been battling the aridity. In the incubator, there's a giant saucer, the kind you put under potted plants, that I've filled to the brim with water. To this, I now add two large kitchen sponges to increase the surface area for evaporation. I'm relieved when, after a while, the humidity holds steady at 82 percent.

I can no longer touch the eggs or open the incubator. The goslings are now completely on their own. I can't help them hatch, and there's no physical way to make the process any easier. All I can do is watch and wait to see what happens—a rare feeling of powerlessness for me. As I sit in front of the eggs, I let my thoughts wander. I've always been interested in animals, but I've never been drawn to avoiding them in my diet, and I don't consider myself to be especially fond of them or overly obsessed.

I let my thoughts drift, keeping my eyes fixed on the eggs, but they look just the same as they did before. Unable to sit still, I stand up and then sit back down again. How are the little goslings supposed to get out? Aren't the shells much too hard? And how do they even know that there's a world out there beyond their shells? Or perhaps there aren't any goslings inside at all? And does anyone have any heart meds they can hand over?

• • • • • • • • •

TWO DAYS BEFORE THE ESTIMATED DATE OF HATCHING, I'm reading aloud again from *The Wonderful Adventures of Nils*.

"The boy simply could not make himself believe that he had been transformed into an elf. 'It can't be anything but a dream—a queer fancy,' thought he."

Every once in a while, I raise my head to check if my audience—that is to say, the eggs—is paying attention. But, as usual, they don't seem to be taking much interest in the story.

"'If I wait a few moments, I'll surely be turned back into a human being again.' He placed himself before the glass and closed his eyes."

At that moment, something amazing happens. The goslings respond. I hear the first soft peep, and I feel my heart warm. There they are. The first sounds from the geese.

I feel incredibly moved. I squeeze my eyes shut and can't believe what I see when I open them again. Apart from the sounds they're making, the eggs are moving very slightly—fractions of an inch in one direction, then another. Clearly, they recognize my voice. I hardly dare to leave the geese alone in the incubator overnight. On the way home, it hits me: there are now nine tiny creatures in there waiting for me.

And then things get really serious. At seven o'clock in the morning, my phone vibrates. It's one of the wildlife technicians from the Institute and she sounds very excited. While I'm still lifting my phone to my ear, I hear her say, "One egg has cracked!" I'm awake right away. Now my adventure with the geese can begin.

name for a budgie or Rex for a dog. I've no idea why. It's just the way I feel. And on this most joyous occasion, as the first egg is hatching, I find Gloria to be a most fitting name. And so, little goose Gloria will be the first to see the light of day. Or, to be more precise, the light of a ten-watt refrigerator bulb.

• • • • • • • •

GLORIA LOOKS AS THOUGH SHE KEEPS TAKING A TIME-OUT to catch her breath from her task of chipping a circular hole in her shell with her egg tooth. The egg tooth isn't really a tooth at all but rather a pointed horny growth on her bill that will disappear soon after she's hatched. This tooth is incredibly useful and important, because Gloria's bill is still too soft for her to rely on it to crack her shell.

It takes an hour before Gloria finally pokes her whole bill out from the inside the egg. I still can't see her head, and for a while after that nothing more happens. But then everything moves rapidly. Gloria pushes rhythmically against the egg wall with all her might, arches her back, and suddenly her whole body is out. Her neck is already incredibly long, and she is sticky and damp. She looks exhausted and a bit forlorn.

"Weeweeweewee. Gagagaga," I trill, making sounds I often used when I was talking to the eggs while they were incubating.

Normally, the eggs would be hatching under mama goose's bottom. Instead of squatting next to them, eyes damp with emotion, cheering the eggs on while they're hatching, mama goose shows her excitement by sitting on her little ones and keeping them nice and warm.

Gloria tries to lift her head in my direction, but she's still too weak. So she just lies there, among the eggs, which don't yet know anything of the world, and recuperates in the comforting warmth of the incubator.

Gloria must now lie for about another twelve hours in the warm, damp atmosphere of the incubator to gather her strength and dry out. And that's also how long it will take for her downy feathers to be released from their keratin sheaths. You can think of these tiny sheaths like small parchment scrolls in which the down feathers are rolled up and well protected like precious arrows. The sheaths look as though they're made of delicate paper, but they're actually made of keratin. While the gosling is drying, the keratin disintegrates, and one by one, the downy feathers unfurl. The result is the light brown to golden yellow fluffy feather coating of a freshly hatched gosling.

• • • • • • • • •

LEAVING GLORIA TO DRY IN PEACE, I SETTLE DOWN WITH a coffee on the Institute's rooftop terrace and gaze out into the green space beyond. It's already unseasonably warm for spring. If I lean forward a bit, I can see the row of large aviaries stretching away from the Institute's main building. This is where researchers are studying blackbirds, carrier pigeons, and many other birds. These researchers, however, don't name their charges, because their research doesn't include their birds getting to know them.

At the far end of the aviaries, idyllic and isolated in a large field near the rear of the property, is the camping trailer. It's right up against the wood. That's where I'm going to be living with the geese for the next few weeks. For our small family, this is going to be our home away from home at the cushy end of town. With its newly appointed nursery, I think of it as our very own Duckingham Palace.

Thoughts are rattling around in my head. What else do I have to do? Am I really adequately prepared? Right now, Gloria is lying in the incubator just as she should be. She doesn't need me yet, but in a few hours, she's going to be completely dependent on me. Then I'll be a full-time single father goose, and unlike a mother goose, who only incubates four to six eggs at a time, with up to nine goslings, I'm going to have a particularly large family to care for. And I'll be goose and gander all rolled into one.

After the geese hatch, something as simple as going out to have a beer without them—or just being alone—will no longer be possible. In my case, there isn't a mama goose who could take over for me. Many human parents (and most geese) have that option. And there's no way the little ones are going to accept a babysitter. How long do I have left to enjoy my freedom without the goslings? What is it exactly that I absolutely still have to do? Why exactly do I need this freedom?

WHEN I RETURN TO THE INCUBATION ROOM THREE HOURS later, I see that two more eggs show signs of pecking from the inside. Gloria is still lying there in the middle, exhausted. I can now hear quiet peeps coming from many of the eggs. The sound is like an eraser rubbed over metal—not so much a peeping as a quiet squeaking.

It's no accident that the eggs don't all hatch at once. If they did, the mother goose would have to care for four to six goslings all at the same time, and that would be too much for her. It's even thought that the goslings peep to get their act together and establish a hatching order. That way, the parent geese have enough time to give each baby goose the attention and care that are so important in the first hours of its life. I imagine one human twin signaling to the other while still in the womb, "Hey, you. Be my guest. I'm not in any hurry."

• • • • • • • • •

I LEAVE THE INCUBATION ROOM FOR ANOTHER COUPLE OF hours and look over the camping trailer one more time. It's actually not that luxurious. I suddenly get cold feet. What if I get cabin fever? Will the geese get on my nerves? Will I get bored with them? Will I really be able to replace a mother goose?

When I finally return to the eggs, my goose family has grown to three. Two more goslings lie matted and exhausted in the incubator. Gloria, however, doesn't look bedraggled anymore. She's dried out and standing on the hatching drawer, a golden fluff ball of downy feathers. She greets me with a loud peeping, as though she's trying to tell me something.

Can't you see me? I was born first! Take me home! Come on, what are you waiting for?

Her down coat is sticking out a bit, which makes her look even odder, as though she has accidentally come into contact with an electrical socket. The filaments of her downy feathers are indeed electrically charged, but that's quite normal and serves a useful purpose. It ensures that the feathers keep their distance from each other. The distance creates air pockets between them, which increase the feathers' insulating effect.

Oxytocin floods my system when I open the door to the incubator and carefully lift Gloria out using both hands. Her down coat is just as cuddly and soft as it looks, but her tiny body feels fragile as I hold her. She peeps nervously, and I can feel her heart beat. The first physical contact is critical to the whole imprinting process. If Gloria is to accept me as her goose papa, she has to be close to me from now on and soak me up with all her senses. This is the defining moment, when the connection between the little goose and me must be established. I bring Gloria up to my mouth and whisper to her as though she were a human baby.

"Gagagaga," I say. "Weeweeweewee, weeweeweewee. Who's this little one, then?"

I'm happy there's no one around to see me, drenched in oxytocin as I am.

TO ME, GLORIA LOOKS AS THOUGH SHE HAS ALWAYS BEEN Gloria, but she also looks like any normal gosling. And, as we all know, goslings tend to look alike. Despite having fallen head over heels for her, I doubt I'll be able to tell Gloria apart from the other goslings in a couple of hours. Although I'm a proud papa goose, I'm no magical goose whisperer. And so I've decided to use colored bands so I can tell the goslings apart. Gloria gets a pink band, which I carefully pull up over her foot. It's quite loose right now, but her feet and legs will grow very quickly.

Normally, this would be the time for Gloria to slip under her mother's feathers. That's not going to work with me, of course, even though for a brief moment at the outset of this project I considered creating a goose rump for myself by attaching a few feathers around my hips. Instead, I just pick her up and tuck her under my wool sweater, directly against my skin. That way, she'll get my body warmth and smell—and I hope that even though I'm wearing wool, she won't think I'm a sheep.

With this little peeping fluff ball under my sweater, I walk along the edge of the wood back to the camping trailer. Unfortunately, that's when it occurs to me that I've completely forgotten to stock up on anything to eat or drink. But I have remembered to fill the yellow laundry basket with wood shavings and place it next to my bed in the trailer like a makeshift cot. The idea is that while I stretch out on the bed, Gloria will sleep on the floor next to me.

"Look, Gloria. This is where we live now," I tell her as I put her in the laundry-basket-goose-bed.

I just want to quickly brush my teeth before I lie down next to her.

"Papa's just going to get ready for bed," I whisper.

Fat chance. My movement away from her is immediately met with heartrending peeping. Gloria doesn't want to stay in her bed alone. A familiar feeling comes over me. It's amazing how this peeping gets to me. After all, it's only a baby goose making a bit of noise, a member of a different species that doesn't really have anything to do with my daily life.

But I'm acutely aware that the little goose needs me as though she were my own baby. She yearns for me. She wants me to be there for her. I must be there for her. And right now she's peeping for just one reason: I moved a couple of steps away from her for a couple of seconds. Here's a wild animal as dependent on a human being as she would be on her own mother or father. This is definitely different.

I shove Gloria back under my wool sweater and hastily brush my teeth. Then I try again. I place her in the laundry-basket-goose-bed, lie down on my bed next to her, and turn out the light.

"We're going to sleep now, my little goose," I whisper.

But Gloria doesn't like being so far away from me. She peeps and fusses to let me know I might as well forget my plan. What will it take to train her to sleep in her own bed? Do I even have the right to make the rules as to how a different species should be raised? I sigh, bend down, and lift Gloria onto a soft towel next to my head. The peeping stops immediately. Less than five seconds later, Gloria slips from the towel and under the covers. She cuddles up on my chest and is completely quiet. I think she dozes off right away. There's no way for me to fall asleep now. I'm far too afraid of crushing the tiny gosling if I move in my sleep.

As I lie there, I can't get what has just happened out of my mind. Mostly, I'm amazed at how quickly the tiny gosling has found her way into my heart. We've only known each other for a few hours. I try to figure it out. Perhaps it's because she so obviously needs me. That's something new and unexpected for a member of the modern internet age like me. We're used to being the ones who need nature and the wild animals that live there. They're supposed to supply us with all kinds of nourishment. They're supposed to feed us or appear in heartwarming YouTube videos. The idea that it might be the other way around and animals might need us is less familiar, yet that is the only way the relationship between people and animals can be truly complete. Just as we need animals, they, too, need us.

AT QUARTER PAST FIVE I'M AWAKENED BY A LOUD "Cuckoo!"

What kind of alarm is that and who brought the damn cuckoo clock? I wonder, still drunk with sleep. I open my eyes wide as soon as it dawns on me. The cuckoo is our next-door neighbor and very much alive. He moved into a tree next to the camping trailer long before we arrived, and I believe he lives there with his wife. I noticed both birds earlier, but it never registered that they might rouse me from sleep with their calls.

If you notice this small bird while you're out and about during the day, you might stop briefly to admire it, but by evening you'll have forgotten all about it. It's only when you're living right next to it, and with animals, that the presence of the cuckoo really registers. It checks in early each morning, and then, in the way of all animals, it goes about its daily business, living a life of which we're usually blissfully ignorant.

Unlike me, the golden fluff ball squatting on my chest and peeping contentedly is more than happy with the cuckoo. I swear Gloria's got something to say.

Isn't it great, Papa? I was awake anyway!

My bones and my neck feel locked in place, probably because I didn't move an inch all night for fear of harming the gosling. Also, my chest is covered in baby goose poop.

Gloria, however, seems totally refreshed and is already exploring the bed. I have to be careful she doesn't fall off. Right now, her tiny bill is still relatively dark. Later, when she's mature, it'll turn reddish orange. A goose's bill is both its nose and its mouth, and the two nostrils on little Gloria's shiny black bill make her look particularly adorable. I can still see the egg tooth at the tip. It looks a bit like a yellowish pimple.

I rub my eyes. I want to go into the bathroom for a couple of minutes, so I deposit her carefully in the laundry basket. She promptly responds with her pitiful cries at being abandoned, "Eeek, eeek, eeek." There's absolutely no way I can ignore this noise. It has the same effect on the levels of adrenaline, cortisol, and oxytocin in my bloodstream as the crying of a newborn baby has on the hormone levels in a brand-new mom.

To be fair, even a brief absence is no small thing for such a tiny gosling. For some time now, research has shown that goslings whose parents—whether they are human or bird—leave them alone too long can develop behavioral problems later in life. I have, therefore, no choice but to accompany Gloria on her first exploration of the landscape of the camping trailer. It's unbelievable how much energy is packed into this tiny bird, and how happy and alert she is as she eagerly sets out on her voyage of discovery.

There are so many things that she's never seen before. She rummages through my socks with her tiny bill. She ought to be familiar with that smell by now. Then she nibbles at the plug for my laptop charger. She investigates everything at length. Then, suddenly, she is exhausted. She glances beseechingly in my direction and begins to make a wavering whistling sound.

"Weeweeweewee," she trills.

This is the goslings' so-called sleepy call. It sounds roughly as though someone's blowing softly into a whistle, gently moving the pea inside back and forth. This is how the goslings let their parents know they're tired and need to rest for a while. So what Gloria is softly whistling to me is something like this:

Papa, lie down and let me crawl under your sweater, but then please hold still. I'm still so small. You can't possibly expect me to sleep alone in that awful laundry basket.

"But we have to go and see the other goslings," I object.

But Gloria has already decided. She crawls back under the covers and dozes for another half an hour while I, unable to sleep, carefully shake and move first one arm and then the other and then both my legs so that they don't go completely numb. I'm already feeling stiff all over. What's it going to be like when nine goslings want to cuddle up in bed with me?

3

Seven Goslings

THE NEXT DAY, GLORIA GETS SIX BROTHERS AND SISTERS. Four have already hatched, and the other two have begun to peck at their shells. There are just two eggs without any sign of activity yet.

Naturally, when you have that many children, you can't give them all the attention you gave the first one. Even so, I delight in every featherlight, fluffy gosling as I take it out of the incubator. But gradually, I get habituated to the process, just as parents must do after the arrival of their second, third, or fourth child. Perhaps one day Gloria will become the petulant big sister who, even in adulthood, will suffer from the fact that she's no longer the center of attention like she was that first night.

I place the laundry basket next to the incubator and attach a warming lamp to the side. So that they all imprint on me, I hold each and every one of the goslings for a while before putting them into the basket. Then they can go right ahead and cuddle up to their siblings before they get cold.

"Greetings," I say. "You're number four."

"And weeweeweewee to you, too."

I'm finding it a bit confusing to have so many goslings. How do parents of triplets or quadruplets manage?

My presence and devotion are vital for the goslings right now in these first few moments of their lives. Imagine the imprinting process as something like stamping a coin. Once the coin has been stamped, the process is irreversible. Once a gosling has seen and smelled me after hatching, it retains the imprint of me as father goose for the rest of its life. You could say that my image now appears on the papa-coin of each goose. Let's hear it for Goose Michael as president!

By now, there's quite a babble of peeping in the incubation room. I'm convinced that the goslings peep back particularly vigorously at the sound of my voice. And why wouldn't they? They've been hearing my voice from behind the walls of their shells for long enough. Now they get to see what the guy who's been trying their patience all this time with Nils's story looks like, as well.

As it hatches, a gosling takes a big leap out of its egg, shakes itself, and looks at me. It still has a little piece of eggshell stuck to its head, which looks very funny. I decide to call this gosling Calimero after the Italian cartoon character—a small chick that always runs around with half an eggshell on its head.

As I greet each gosling individually, I have time to think a bit more about their names. Calimero and Gloria are a perfect fit. After that, I simply work my way down the list my daughter gave me. And so, the next five baby geese are named Nemo, Maddin, Frieda, Paula, and Nils. Each gets a band of a different color. And so I don't immediately forget which is which, I carefully write the names down in colors that match.

I name the goslings, even though I don't yet know which sex they are. I could find out by examining their cloacas. In geese, that's the name for the single opening where both the urethra and anus come out. But that would be very unpleasant and painful for the goslings. I don't want to hurt them. It'll be a few weeks before I find out which are geese and which are ganders.

The whole process of hatching, drying, and imprinting proceeds fairly smoothly. It's almost as though we've all done this before. I was so stressed and worried beforehand that I'm actually quite surprised. It was all much easier than I thought it would be. Looking at me with their little obsidian eyes, the geese seem to find such levels of stress and doubt unfathomable. They're simply here, hatched and ready. And now? There's a whole world to explore. I think I can read their thoughts.

Can you believe that idiot thought we might not make it?

Why do you foolish humans always worry so much? Have a little more faith in nature.

FOR NOW, I LEAVE NILS AND FRIEDA IN THE INCUBATOR. The rest of the flock is coming home with me tonight. I bring the peeping laundry basket into the camping trailer. A door-to-door stork delivery service? No, make that a door-to-door goose delivery service.

"Welcome to your hotel, custom designed for its unique clientele," I say.

The goslings peep. Whether they're agreeing or disagreeing, I can't quite decide, but I can't imagine that the accommodations will disappoint. Although I have to say that, even though I've gone to a lot of trouble with the camping trailer in the past few weeks, the setup is still fairly Spartan. The only nods to my personal comfort are a small bed, a minuscule bathroom unit, and an electric refrigerator.

In case of inclement weather, I've built a small deck out of wooden pallets in front of the camping trailer. I've put a sheet of plywood over them with a small folding table and a couple of folding benches like the ones used in outdoor cafés and beer gardens. If it rains I can sit comfortably under the cover of the trailer's awning while the goslings play outside. At least that's my plan.

The camping trailer is pulled up next to one of the Institute's aviaries. I've taken out one of the aviary's side panels so that the structure is now practically part of the trailer. The idea is that I will be able to see and talk to the geese from my bedroom window. What I really want is for the geese to sleep in the aviary at night so that I'll have my bed to myself—for a few hours, at least.

There's already a "gosling corral" inside the aviary, custom-made by one of our wildlife technicians. In these early days, it's important that there are no right angles in the goslings' quarters to ensure that none of the little ones gets inadvertently trapped in a corner by the others, where it might suffocate. The goslings look fairly well developed, but their bodies are still delicate and easily damaged. If I held them too tightly, I could crush them.

The corral encompasses about twenty square feet and is made of flexible glazed pressboard a bit like the material used for the back of Ikea cupboards. It's tall enough that the goslings won't be able to reach the top, even when they're bigger and can jump. I've attached two infrared lamps and a selection of food and water dispensers to the side.

I think all these amenities must merit at least four feathers in the goose resort rating system. Of course, with their colored leg bands, the baby geese have booked an all-inclusive stay with daily programs and excursions.

• • • • • • • • •

AS SOON AS WE ENTER THE CAMPING TRAILER, THE FIRST thing to do is have a good cuddle. It feels wonderful because the goslings are so soft, but it also means that I absolutely have to change the top part of my ensemble. The goslings are world-class food processors, and they all poop at least once before I manage to get them out from under my sweater. Alas, geese aren't like cats. You can get them to accept a person as their mom, but no one has ever managed to train them to use their own goose litter box.

Goose droppings are relatively solid and don't have much of a smell, so I'm not overly grossed out, but I prefer not to think about the quantity of droppings seven grown geese will soon be depositing daily around our camping trailer. Thankfully, greylag geese are vegetarians. If I were surrounded by other birds—blackbirds, for example, which eat meat as well (worms mostly) and therefore produce completely different droppings—the smell really would be unbearable.

After I put the laundry basket down and turn on the warming lamp, I collapse like a dead weight onto my bed. Unfortunately, less than two minutes later, the young geese begin to call for me, even though I'm less than ten feet away. Their papa needs to stay with them.

"Weeweeweewee!"

They quiet down only when at least one of my hands is in the laundry basket. I can forget my beautiful plan of putting them in the gosling corral in the aviary overnight. However, they can't all come into bed with me, either. After some back and forth, various positions, and a lot of peeping, I decide on the following arrangement: I lie on my bed on my stomach with one hand in the laundry basket, as though I'm sleeping with my hand in a baby bassinet.

It's all very cozy as the goslings press up against my hand, but I know all too well that at some point my hand is going to go to sleep and hurt like the devil. And so, as a precaution, I rub it with pain-relieving gel, which doesn't bother the goslings at all. The main thing for them is that I'm with them—or at least part of me is. Besides that, I pacify them by talking to them in "goose." My first night with five goslings can begin.

• • • • • • • • •

IN THE HALF-DARKNESS, I WATCH THE GOSLINGS AS THEY cuddle up against each other in the laundry basket. True, from time to time, they push each other away from the grain dispenser, but then they go back to warming each other as though the shoving match never happened.

There's something soothing about the peeping laundry basket. I'm happy to have these creatures so near me, though I can't really explain why, even to myself. The birds don't seem strange to me at all; indeed, they already feel familiar. As I lie in bed next to the geese, it's as though something's missing in my chest, something that should be there. That's odd.

I certainly didn't come into the world as someone with a soft spot for geese. For Christmas dinner the year before last, I didn't hesitate before shoving a stuffed goose into the oven. Truth be told, the bird tasted delicious. And yet, here I am, lying in bed in a camping trailer with the baby geese, and I know this beyond a shadow of a doubt: these goslings need me as unconditionally and definitively as a tiny human baby needs its parents.

The goslings' presence both moves and grounds me. The goslings know only simple, basic things: how to eat, explore, defecate, rest, and sleep. They don't want to do anything else, and they don't understand anything else, either. All those things that normally stress me out—the expectations of others, financial matters, relationships, obligations, shopping, etc.—simply don't mean anything to them. Perhaps, I think, as my hand is already beginning to hurt, love doesn't necessarily have to be between two people but is something that has to do with our existence and sense of belonging in nature in general. Then I suddenly see a golden yellow dog, me absolutely alone in the air, and a number of geese dangerously close to the plane propeller. And then I, too, fall asleep.

• • • • • • • • •

I'M WOKEN BY UNBELIEVABLY LOUD PROFESSIONS OF LOVE. It's Mr. and Mrs. Cuckoo exchanging adoring messages from their respective trees. Can't they keep their feelings to themselves or at the very least perch on the same tree and whisper their sweet nothings to each other instead of yelling? What I'd really like to do is grab the heavy-duty hose and treat both of them to a morning shower.

"What do you think you are? Two roosters?" I shout outside, somewhat annoyed.

Not that I have anything against cuckoos in principle. But can't they just sleep a little longer? Is that too much to ask?

The cuckoos' calls clearly signal to the goslings that it's time to party. They peep excitedly, impatiently demanding my attention. It's only quarter past five, but I basically have no choice but to get up. I learned this when my own children were still small. I get dressed, shove a camping mat under my arm, and march out of the camping trailer and onto the field. And what about the goslings? They instinctively goose-step their way after me. In single file, no less. They can do it and they know how to do it, even though no one has shown them. Not me, anyway.

I'm happy that the imprinting on me has worked so well. The geese follow me in a picture-perfect goose-step march. None of the birds makes a move to separate from the group and wander through the field alone—a very sensible and lifesaving instinct. It's a clever move on nature's part, because getting separated from the group would spell almost certain death for an individual in the wild.

Outside, the goslings discover the taste of fresh grass and let off steam while I sit on the camping mat. The goslings aren't yet strong enough to use their bills to rip off big tough blades of grass. But, no matter, they soon will be, and they can survive for another couple of days on the yolk sac they absorbed when they hatched.

They test the grass anyway and give part of the field a good fertilizing while they're at it. Then, suddenly, they're tired. They all begin to whistle softly and crawl up under my sweater. With the goslings close to my chest—they feel a bit like a heated feather pillow with a mind of its own—I make my way back to the basement of the Institute.

"We've got to go and collect the other two rascals," I explain.

And, indeed, Frieda and Nils greet me with vociferous peeping and are more than happy to join the others under my sweater. The goslings on my chest are beginning to feel rather strange. My whole sweater is peeping and wobbling, and I feel a bit like a magician who's about to make a gosling appear out of each sleeve.

I ascertain that the last two eggs show no changes at all. I listen and watch them for half an hour to see if I can catch the slightest sign of life. But all is still inside the eggs. And so it is with a heavy heart that I decide to take the eggs out of the hatching drawer and put them in the aviary compost pile. There's nothing else I can do. If the goslings haven't hatched out of their eggs by now, they won't hatch in the next few days, either, and I can't help them by chipping the eggs open from outside. Unfortunately, it's not unusual for a certain percentage of goslings to fail to hatch and die while they're still inside their eggs.

• • • • • • • • •

THAT AFTERNOON, I PUT A SMALL KIDDIE POOL INTO THE aviary, and I'm surprised at how excited the goslings get. A credit to his name, Nemo is the first to jump boldly into the water, and he immediately dips his head under. I've no idea how he's learned to do this. He can simply do it, even though he's barely two days old.

Nemo doesn't need to learn how to swim or gradually get used to this watery element. Nor do I have to explain to him how he needs to move in the water, which is different from how he moves on land. He just jumps in. It's not long before all seven baby geese are happily frolicking, splashing, and playing submarine. It's quite astonishing when I think that just a short while ago these fun-loving explorers were still motionless white eggs.

On the second night, the goslings are meant to sleep alone in their gosling corral in the aviary for the first time. I put one of my T-shirts down on the wood shavings as a sleeping pad so that they won't miss me too much. I've been wearing it almost nonstop for the past few days, so it smells unmistakably of Michael's Musk. I check the temperature of the ground outside and hang a second warming lamp in case the first one stops working. I don't even want to think about what might happen if the little ones were to go a whole night without heat.

"There you go," I say. "You can sleep over here. Weeweeweewee."

At first, the young geese seem to be really interested when I put them in their new nursery. They look around inquisitively. I take this opportunity to carefully withdraw. Immediately, I'm reminded of the times when my children were still little and how, when they finally fell asleep, I tried to steal out of the room on tiptoe.

Of course, less than five minutes later, the goslings are protesting loudly.

"Weeweeweewee," they cry.

But this time I don't let them get to me. After all, a bit of discipline can't hurt. Instead of getting up from my spot in the camping trailer, I just look out the window and jabber "weeweeweewee" in the most calming tone I can. My plan works: when they see my face and hear the sound of my voice, the geese stop calling.

There's just one problem. How am I going to get myself something to eat? Jabbering back and forth with the geese has made me quite hungry. I ponder this for a while and imagine how a pizza delivery person might react to my accommodations and the geese. Probably it wouldn't be a big deal. People in that job must see just about everything. Or perhaps, to really be a papa goose, I should try the grain the geese seem to be enjoying so much?

Even though there's no point, I open the tiny refrigerator again—and step back in delight. Our caretaker at the Institute, Heinrich, has left me a large cheese sandwich with a note that says, "For Goose Michael. So you don't feel you have to eat the geese." What a stroke of luck! Now I can relax. I grab the sandwich and tiptoe outside to my homemade deck.

There's still not a cloud in the sky and it's absolutely quiet out here. The sun sinks slowly behind the wood, the oats in the field opposite are bathed in golden light, and a contented, peaceful feeling gradually soaks through my brain. It feels as though everything is right with the world. This is an emotion I haven't felt in a long time.

I'm aware that we've passed our first big milestone. The baby geese hatched out of their eggs without any great incident. Everything went smoothly. Apparently, nature can manage most things quite nicely on its own. Before I become overly sentimental, I reach for a rake and a shovel. Without giving it a second thought, my handful of goslings have deposited their body weight in goose droppings in my front yard. Of course, they don't care who has to clean this up. And why should they? The caretaker with the wooly sweater in a lovely shade of goose-dropping green is always around, isn't he?

4

Our First Dip

TO SOMEONE OUT FOR A WALK AND OBSERVING THE SCENE from a distance, it probably looks ridiculous. There's a grown man walking along a path in the meadow, gesticulating, honking a horn for no reason, calling out random names of cartoon characters, all while making odd noises.

"Calimero! Weeweeweewee! Nemo! Gagagaga!"

But it's just me with the little geese. We're walking quite normally along the path—in the way that geese do. We're out on our first long walk, and I want to show the geese something of the neighborhood. That's an important task for a mother goose: to give her children an overview of the place where they live. This exploration is much more important for a mother goose than it is for our little party. There are dangers and obstacles everywhere for her that I can do something to protect the goslings from: roads, people, dogs, houses, and all kinds of vehicles. On their first exploration, the mother goose shows her babies where they are safe, where they must pay attention, and where they must never, ever go.

"Geese," I announce. "We have arrived."

I've already become used to talking to the geese as though it's nothing out of the ordinary. For example, I give them a running commentary on what I'm doing and why, as you do with a pet or a baby. I know they can't understand me, of course, but if I didn't say anything, it would be even more peculiar. If I didn't talk to the geese and I wanted to get them to do something, I'd have to wave my hands about as though we were all in a game of charades.

The goslings waddle after me, and as they do, they sometimes extend their fluffy, stubby wings, which looks comical and cute at the same time. My original plan had been to waddle among them like an oversized goose, but I quickly abandoned that idea after I almost fell flat on my face in the mud. And anyway, I don't want to cozy up to the goslings too much, as my goal is to be a convincing, if human, father goose. And that's why I'm just sauntering along ahead of the baby geese and enjoying their amusing company, as though walking along like this was the most natural thing in the world.

I find them to be so attentive and trusting. There's always something for them to see, and they find everything interesting: individual blades of grass, the columns of ants that cross our path, the reeds beginning to grow alongside the path, the gravel that looks like dark kernels of corn. All these distractions mean that they don't always keep up with me as they waddle but fall behind from time to time, flapping their stubby little wings as they rush to catch up.

We walk and waddle along a small track still damp from the recent rain that leads to an idyllic lake with a swimming platform. To our left lies a field of corn; to our right an overgrown grassy meadow. In summer, the meadow will be red with poppies and abuzz with insects, as though it were a gigantic market square laid out just for them. The lake is about half a mile from our camping trailer. People seldom swim here, especially not in this unsettled weather, so we will be relatively undisturbed. For our first attempt at swimming, I'd like as little distraction as possible.

The first thing to do, however, is get to the lake on foot—webbed or otherwise. I walk slowly backward so that I can be in front of the geese and watch them at the same time. The little goslings look pretty much alike—they're yellow and fluffy. But I'm already noticing ways in which they differ and prefer different things.

Nemo, for example, has become a real water baby. Perhaps he imprinted on his name as well as on me. Yesterday, he was the first to put his head underwater in the kiddie pool, and he stayed in the longest. Even now he's obsessed with every puddle, no matter how small. He takes a running jump into all of them. When we continue without giving him enough time for the puddles, he peeps and is thrilled when we wait. *Just like my son*, I think. There was a time when he couldn't pass up a puddle, either—whether he was wearing his rubber boots or not.

There's something else Nemo likes. He's taken to being my dentist on the side. When I lie in the grass with the goslings, he's especially keen on using his bill to inspect the inside of my mouth. I don't know if he thinks I've got seeds stuffed in my cheeks like a hamster, but I don't particularly enjoy his examinations. Thank goodness he doesn't see the need to drill.

In contrast to Nemo, Calimero's turning out to not be such a good fit with his somewhat goofy namesake, either in character or appearance—he's not black like the little Italian cartoon duck, for starters—but perhaps he thinks he's still wearing that eggshell helmet and this makes him invincible. Whatever the reason, he's the most daring of the bunch. He shrugs off collisions with the others. In fact, he seems to pointedly seek them out. Perhaps Rambo would have been a more appropriate name. Anyway, Calimero is aware of his Rambo-like qualities, and with Nemo, he has taken on the role of watchdog and protector. Nemo and Calimero defend the camping trailer against intruders. When Heinrich dropped by just now with another sandwich, the pair rushed at him like tiny *Tyrannosaurus*-geese. Heads held low, they cornered the intruder and tore at his pant legs. Heinrich just managed to save himself with an energetic jump over the chain link fence.

"What kind of killer geese are you raising there?" he shouted to me. But then he had to laugh.

Unlike people, geese do not let others do their dirty work for them. As leader of the group, Nemo has no problem being the first to stick his bill into the fray.

So far, Paula and Nils are lowest on the social ladder. Rather than pushing them into the line of fire, Nemo protects them. Although that does mean that when it comes to swim time, they're the last ones allowed in the kiddie pool. Paula and Nils don't have much say in the group, but that also means that life on the lower rungs of the social ladder runs relatively smoothly. They don't have to defend their lowly positions against the others—as the bosses Nemo, Calimero, and Gloria must do—and therefore they have plenty of time to interact with me. Whenever Calimero and Nemo begin peeping at each other, Paula simply turns to me. She trusts me completely, and if it were up to her, we'd spend the whole day cuddling.

Frieda is the most stubborn of the geese. She's not at all interested in the puddles along the way and has no desire to splash around in them with Nemo. She's far more interested in getting her own way. She's a loner and often distances herself from the other geese.

The geese behave very differently, so it makes sense to me to talk about them having personalities. How is it that we so rarely notice this? How is it that we talk of "the geese" or "the chickens," as though the birds were all the same? Perhaps a stubborn, ornery goose like Frieda has more in common with a stubborn, ornery chicken than with a well-behaved, obedient representative of her own species.

While I'm mulling this over, I notice that Frieda has once again dropped out of formation. She's waddling, as calmly as you please, away from the other geese toward the meadow. Here's an opportunity for me to experiment with my horn, which I've been holding in my hand all this time. I give the black bulb a hefty squeeze, and the sound blasts out over the field—to dramatic effect. Frieda lifts her head right away, races up to us—wing stubs raised—like a streak of goose lightning and immediately rejoins the group. The imprinting with the little horn has apparently worked like a charm.

I honk the horn a few more times during the walk, mostly for Frieda, who simply will not fall in line. Each time, order is immediately restored to my goose family. No sooner does the horn sound out over the field than the little geese all rush to me. What a shame, I think, that there isn't something similar for small children. Then all you would have to do is play a sound repeatedly to the mother-to-be's baby bump to have control over the wildest hellions later in life. I just hope that the imprinting has worked as well with the sound of the propeller as it has with the horn.

I'm amazed that the geese don't make a run for it while we're on the path. They have plenty of opportunities, despite my horn. The grass is already so high that if they were to waddle just ten feet or so away, I wouldn't be able to find them. Rather than standing out from the undergrowth like a bright neon yellow sign, their yellowish brown downy coats blend right in. A couple of goslings could quite simply decide to walk off in opposite directions at the same time, and I would have little chance of catching them both—and no chance at all if four goslings decided to scatter. But not a single gosling makes a serious attempt to leave.

The baby geese follow me of their own free will. They seem to trust me and feel that they belong to me. I don't have to constantly encourage them to keep walking, either. They seem to enjoy moving for the sake of moving and show no signs of wanting to throw themselves screaming on the ground like obstreperous children. Not a single one of them whimpers, *Papa, weeweeweewee, how much longer? Are we there yet?*

Seven Things Goslings Can Do on Their Own

1. March in single file
2. Peep for their papa and tug at his heartstrings
3. Hop into a kiddie pool
4. Look out for each other
5. Recognize the meadow with the tastiest grass
6. Whistle in their sleep
7. Cuddle up in a way that is particularly uncomfortable for Papa

"WELL, HERE WE ARE," I SAY, WHEN WE FINALLY ARRIVE AT the swimming platform.

I'm pretty sure I hear an excited peeping when they see the water. The lake isn't particularly large, but for such tiny little geese, who so far have only set eyes on a kiddie pool, the amount of water has to be quite an impressive sight. The water temperature soon gives me something to think about, too. It's barely fifty-seven degrees Fahrenheit. While the geese break for a quick snack of grass at the edge of the lake, I squeeze myself into my full-body neoprene wetsuit. It seems to me as if the geese are already pawing the ground with their little webbed feet, because they can hardly wait to finally swim in a real lake.

Hurry up, old man. We want to get in!

What kind of a skin is that he's putting on?

Wasn't that seal over there our papa just a moment ago?

Before I've put a foot in the water, Nemo is already in. He just jumps in and swims back and forth close to shore. The other goslings, however, are not quite as daring. They wait in a huddle on the shoreline until the worthy gentleman that is their papa tentatively takes his first step into the cold water. Despite my protective rubber casing, that takes some effort on my part.

No sooner am I in than Nemo's siblings throw themselves in like wild things and begin to charge about, splashing and diving, delighting in the whole experience. They stay close to me, nudging me constantly with their wet bills. My head is surrounded by seven gabbling feather balls, all thoroughly enjoying themselves. The little ones experiment in their new element, beating their stubby wings, and puffing and blowing through their tiny bills. I haven't taught them how to do any of this. How naïve it was of me to think that my first onerous task would be to teach the geese how to swim. All I have to do is accompany them. I'm here to show them that swimming is possible. For the geese, trying to do something and being able to do it are almost the same thing. But they need to have the confidence to try.

Nemo's busy nibbling on my neoprene shoulder when it occurs to me that the geese shouldn't get too wet on their first outing, because there's one thing I'm missing as papa goose: I definitely don't have a preen gland. This gland is especially important for many birds, including geese. It's found on the lower back, just in front of the spot where the tail feathers start. It secretes an oily substance that birds collect on their bills and distribute over their feathers when they clean themselves. Thanks to this gland, they can oil their feathers daily to keep them waterproof. It's because of this constant application of oil that drops of water pearl so beautifully off the backs of geese.

The small goslings already possess a rudimentary version of this gland, but it isn't fully functional yet. Even so, they've already mastered the moves they'll need: they guide their little bills to the spot where the as-yet nonfunctional gland sits and go through the motions of combing the secretion through their feathers. That's another thing, of course, that I haven't shown them how to do. How could I? I'd never be able to reach my backside with my bill. The behavior, it seems, is genetically preprogrammed.

In the wild, the goslings' feathers get a coating of oil when the baby geese slip under their mother's rump. When the goslings were in the kiddie pool, I tried to replicate this effect using my wool sweater, without success, because the water isn't yet rolling off the goslings the way it should. That's why I have to be careful at first with longer swims. If the goslings get too wet, they could lose too much body heat despite their downy feathers. And so, without waiting any longer, I climb out of the water and break off our little swimming excursion.

The geese follow me out onto the shore without complaining. Once I'm on dry land, I honk the horn, though that really isn't necessary. As they shake the water droplets from their feathers, they look as though someone flipped a switch and turned their motors on. On the way back, the goslings look particularly soft and fuzzy, like little kids coming home from the swimming pool, blown dry and all tuckered out.

"So, how did weeweeweewee like the lake, geese?" I ask, and I have to laugh at myself. "Let's do that again soon, shall we?"

I ask them a few more questions, but of course, the geese don't respond. They just waddle behind me in an orderly fashion and peep to show how happy they are.

5

Too Much Work!

TIME PASSES AND THE GEESE GROW LIKE WEEDS. I WAS amazed at how quickly my own children changed, but the little goslings definitely have them beat. Shortly after hatching, the goslings weighed barely five ounces, but a couple of days later, they'd already tripled their weight. That's as if a baby that weighed six and a half pounds at birth weighed as much as an eighteen-month-old toddler just a week later. Human beings are probably the only creatures on the planet that can afford to take so long growing up. Babies and small children, after all, don't have to worry about things like being eaten by a marten before they hit puberty. Goslings, in contrast, are easy prey for all kinds of predators.

I've been totally immersed in the world of the geese for the past few days, and increasingly I'm beginning to see my surroundings through their eyes. It's been days since I checked my email, and I've lost the sense of urgency for pressing engagements. From the vantage point of the camping trailer, the stress of the city seems far, far away. It's amazing how quickly things have changed, but I have different priorities now: keeping an eye on the geese, resting, cuddling, watching out for the weather, and checking out the grass in the meadow. We make brief excursions every day. We often waddle down to the lake, but we also often just take a turn around the field and along the edge of the wood past the small castle where a very nice baroness lives with her husband. The couple owns a lot of land around here.

As I wander over the land that belongs to the castle, I begin a daily "goose meditation." I focus purely and simply on the little goslings with nothing but their needs on my mind. As soon as I do that, I experience a curious soothing sensation of inner freedom. Observing the world through the eyes of the geese, I disengage from myself. Geese see the world very differently from how people see it. Geese don't notice many things that are important and commonplace for people, whereas other things that appear peripheral to us are vitally important for to them.

Seven Things That Don't Interest Geese at All

1. World politics
2. Being cool
3. Shopping

4. Aristocratic titles
5. Whether we go for a walk in the sun or the rain
6. My constant chatter
7. Who cleans up their mess

• • • • • • • • •

WHEN WE'RE OUT AND ABOUT, I ALWAYS CARRY A SMALL backpack containing essentials: the horn, a camping mat, and a snack. I don't need to bring any provisions for the geese, of course, because an ample picnic is laid out for them in every meadow we pass.

As time goes by, I begin to think of the geese as "my kids." I can't really say if that's going overboard or odd. And I really don't care. What's important is that, as the days pass, I'm becoming more relaxed about the way I think about things.

The geese are already so fixated on my horn that I hardly ever have to use it. All I have to do is hold it in my hand and walk away, and they come running from all sides and line up behind me.

I've fallen into the habit of calling, "Come, geese! Come, geese!" while we walk, to remind them where I am as they're getting waylaid by the many distractions along the path: all manner of weeds, flowers, grasses, and bugs. Sometimes I call five or ten times in a row, "Come, geese!" so that it sounds like "Comgeese," as though I were a salesperson for a cellular data carrier: Comgeese—all you need to reconnect with nature.

We often walk by a stream where the geese always want to take a break. They love to include small rest stops along the way to eat, cuddle, and sleep. Nemo is first, of course. With a mighty leap from the small bridge, he jumps into the water below and tears up and down the stream until there's spray flying everywhere. The other small geese love the water as well, but they don't dare jump directly from the bridge, so they first have to find their way down through the underbrush over sticks, touch-me-not, and stones.

The sharp stones the geese walk over hardly bother them at all. It's fascinating how tough the webbing on their feet is. When I touch it, it feels quite delicate and soft, as though it would tear easily. And yet, the geese have no trouble negotiating their way over rough terrain. The tiny claws on their feet also help as they scramble their way down to the stream in single file. Once they get down, the goslings immediately launch themselves into the cold water and swim upstream, using their bills to dabble for roots in the sand.

While they're doing this, I sit on the bridge and dangle my feet in the water. The geese interpret this as an invitation to nibble on my toes. Their bills are now quite strong, which makes some of the nibbling less than pleasant. Geese don't have individual teeth. Instead they have bony ridges set with sharp points like the jagged fortifications on a castle wall. These serrated edges make it easier for them to rip off blades of grass.

As the geese nibble at the weeds around my feet, the differences between their characters come to the fore. Calimero, for example, is a real brute, abusing my big toe as through it were his own personal teething ring. Paula, in contrast, is gentle, nibbling on it very carefully to satisfy her curiosity. Nemo takes no interest in my feet whatsoever and prefers to zoom through the water like a torpedo.

Frieda often stays off to one side, nibbling on tender new growth. Since hatching, she has remained very introverted and wary of me. Although the other geese don't shut her out, she clearly has difficulty completely committing herself to the group. However, she does have one special friend—Maddin—who often seems to act as an intermediary between Frieda and the other geese. Maddin isn't quite as fixated on me as the other geese, either, and doesn't want to cuddle with me all the time. He always looks a bit skeptical, as though he wants to say: *That stupid jerk should just lose the horn. Damn imprinting!*

While I'm musing about the personalities of the geese, I catch sight of Gloria disappearing beyond the next bend. She's drifting away on the current.

"Gloooria!" I call. "Glooooria!"

No response. I grab the horn and honk it loudly three times. Barely five seconds later, I hear the sound of splashing and tiny wings beating the surface of the water. Gloria is powering her way upstream, doing a fine imitation of an adult goose taking off from water. She was probably so absorbed in making all kinds of new discoveries that she didn't even notice how far from the group she'd drifted. Nils greets the runaway by stroking her enthusiastically with his bill—a sign of great affection. He's as happy as I am that "his" Gloria is once more by his side.

Nils is the smallest and lightest baby goose. After hatching, his downy coat was a slightly darker shade of yellow than the others. He never stands out in the group. He's mostly just there. He participates in everything and is particularly close to his eldest sister, Gloria.

It's as important to the geese as it is to me that we are a family, and they interact with each other a great deal. You could almost say that they look out for each other, and the desire to be part of a group has been deeply rooted in them since they hatched. And so our little goose family sometimes sits together for hours by the stream or the lake.

If it were up to the baby geese, we would spend the whole day here in the bushes or in the tangled growth by the water. Sometimes the geese even fall asleep on the water, though normally they prefer to be right next to me. Amazingly enough, where they nap is also part of their genetic programming. They sleep in the middle of the lake only when the current isn't too strong. For a wild goose, drifting toward the shore while napping would be far too dangerous: martens, foxes, and other animals could easily snap up goslings from land.

But even out in the middle of the lake, small greylag goslings are not always safe. A pike might grab them from below and pull a tiny tasty gosling down into the deep. I don't know if there are any pike in our lake, but even so, I like to don my neoprene suit so that I can keep nice and close to the geese while they float, almost motionless, on the surface like adorable super-realistic hunting decoys.

• • • • • • • • •

WHEN WE SET OUT AGAIN IN THE AFTERNOON, WE'RE often greeted by a ruckus from the other side of the stream, although the first time it happened, I didn't even notice. The noise is a clamor of complaint from a pair of ruddy shelducks that are not at all happy with me bringing seven goslings into their little territory. The little geese understood the situation right away and immediately slowed their pace and began to waddle somewhat hesitantly. I, however, had to look up into the sky numerous times before I connected their dithering to the ruddy shelducks.

When they're fully grown, ruddy shelducks are roughly the same size and shape as greylag geese, but their coloring is very different. As their name suggests, their stomachs and backs are rusty red, and they have a few pitch-black feathers in the middle of their backs. Because they scold and complain so much, I have called the pair Toni and Moni.

I don't understand why Toni and Moni get so upset when they spot us. After all, there's more than enough greenery here to feed my seven goslings and any number of ruddy shelduck families. Perhaps Toni feels somewhat superior because of the ducks' awesome exotic plumage. Or perhaps the couple is defending its territory particularly aggressively because they don't feel at home here yet either. Ruddy shelducks are native to the deserts and steppes of Asia. It's only after a few of them escaped captivity in Europe that they expanded their range and settled here.

Anyway, Toni and Moni squawk at my goslings and me relentlessly. It doesn't make any difference if I threaten them with sticks or yell: "Hey, it's only me, Goose Michael." They start scolding us every time we appear. Perhaps they're just both particularly cantankerous representatives of their species.

• • • • • • • • •

THE RUDDY SHELDUCKS, HOWEVER, AREN'T THE ONLY ones eyeing us suspiciously. One day on the way to the lake, we meet a farmer in his tractor. He stops and looks at me and the seven geese, shaking his head. The geese, frightened by the tractor, huddle together between my legs. The farmer opens the window and calls down from his cab in the broad local dialect.

"Wha's thee doin' wi' they geese?"

"I'm taking my baby geese for a walk," I reply, as though that were the most normal thing in the world for a man of my age to be doing.

"They's grey geese?"

"Yes, they are."

"Wha's thee wan' wi' 'em?"

"I'm from the Max Planck Institute."

"Who are thee?"

"I'm from the Max Planck Institute, and later, when these geese can fly, we want to use them to research air currents in the atmosphere."

"They geese 'ave na breast meat an' there's no way t' git 'em real tender."

"True."

"Thee's makin' thysel' too much work! Best buy a pressure cooker," the farmer says.

Then he slams his window shut and races off. I keep waddling with my geese, a forced smile on my face.

• • • • • • • • •

THERE'S AT LEAST ONE NEIGHBOR WHO COULDN'T CARE less about the purpose of our operations and excursions, because all he wants to do is profit from them. This neighbor is Fridolin, a tiny field mouse who regularly uses our walks to help himself to the birds' food and likes to leave the grain dispensers as empty as possible. Fridolin has been getting more and more daring, occasionally even allowing me to catch sight of him while I'm sitting out on my deck next to the camping trailer. Not content just to steal the feed he can eat up right away, the cheeky little fellow is clearly laying up supplies for later.

When I examine the path he has worn into the undergrowth, I wonder where on earth he can have stored such a vast amount of feed. Is he supplying a whole family of mice? Or does it just amuse him to steal from us? I've chased him on many occasions, but little Fridolin has always been too quick for me.

And then there's Jürgen. Jürgen is the dog that belongs to a young woman who walks close by the camping trailer every morning and evening. I've no idea what the woman's name is or why her dog has a person's name, but her dog is not particularly obedient, so she's always calling, "Jüüüürgeeen! Come here for goodness' sake! Jüüüürgeeen!" The woman gets on my nerves almost more than that cuckoo that still wakes me every morning shortly after five. At first, I thought she was calling a drunk boyfriend who had lost his way in the wood or was playing hide-and-seek, but it turns out that Jürgen is a sweet, playful-looking red-brown Irish setter.

When we're out and about, the geese are aware of the dog even before I can hear his owner, and every time, they're terrified. A real mama goose would pick up on her babies' fear of the dog right away and immediately try to get them to safety. But we can't move our camping trailer just because Jürgen takes his walks in the neighborhood. And I certainly can't disappear off the path into the undergrowth when Jürgen's around, as a mother goose would do with her babies, because then I would never be able to find them again.

I briefly considered talking to the woman and asking her to walk somewhere else for a few weeks, but I can't leave the geese alone, and I can't get close to her and her dog with the geese in tow, either. Who knows whether Jürgen would be able to resist the sight of seven flighty geese, no matter how gentle he is. They certainly aren't nearly as delicate as they were right after hatching, but one random snap in their direction would probably be too much, even if all Jürgen wanted to do was play. No matter how well intentioned, a game could easily be fatal. Luckily, I now have a pretty good idea of when the woman walks her dog, and I take care to avoid the path at those times.

• • • • • • • • •

ACTUALLY, THE CUCKOO DOESN'T GET ON MY NERVES THAT much anymore. After five days, we called a truce, even though he does still wake his beloved cuckoo wife with his shrieking at five o'clock in the morning. I'll even go so far as to say that I'm happy he calls at five every morning, because it's particularly lovely outside at that time of the day. People really should get up at daybreak more often and just go and sit outside, but of course, who does that unless they have to?

Early in the morning there's often a veil of mist over the trees, and near the camping trailer it sounds as though the birds in the wood and out in the fields are warming up for a singing competition. That's when the geese lie at my feet, and I sip my morning coffee and observe the goings-on in the neighborhood.

There's a pair of great tits that live in a nearby apple tree. Because they're such traditionalists when it comes to parenting, I give them the solid old German names Ilse and Horst. Ilse spends most of her time sitting on the eggs in nest box number 64, which I put up years ago as part of a great tit research project, while Horst gathers food from morning to night and brings it to her. Horst runs himself ragged for Ilse. Yesterday I counted six incoming flights from Horst in just one hour. By my estimation that means in the twelve hours of daylight at his disposal, Horst must bring food to the nest box as many as seventy times.

While I sit there lost in thought, the geese nibble away at my pant legs, socks, and shoes. They clearly intend to get a reaction out of me and are hoping to make me move my leg or jerk it away. When they succeed, wide grins flash across their faces. That's what it looks like to me anyway. It's as though they're calling: *Papa, papa, let's get a move on! We've been waiting for soooo long! What are you doing?*

"I'm drinking my coffee, you lousy, good-for-nothing geese."

Coffee? How boring!

What they'd really like is for me to get going and take them to the juiciest patch of grass in all of Radolfzell.

"Okay, you annoying little things," I say finally, and we set off for our dandelion field just around the corner.

No sooner do we arrive than dandelion fluff still damp with dew starts flying through the air. In no time at all, the geese eat their fill and then cuddle up to me. The routine is always the same: munch dandelions, cuddle, nibble on Papa, whistle sleepily, and doze off.

As the geese nap, I have time to do a bit more thinking. Since spending time with them, I've noticed an odd change in me. Thanks to the geese, I'm finding it easier to accept that because of my divorce I can't see my own children all the time. It's as though the little geese have gifted me with a generous share of their primal trust in the world. But how can that be? How did a handful of totally normal greylag geese manage to do that?

Although I'm spending all my time looking after the little geese, I find part of the process includes getting to know myself better. That sounds paradoxical, and I think it is. But that's the only way I can describe it. What the geese communicate to me is that things are what they are and it's worth living in the moment.

Isn't that how it works with extreme sports? You give yourself completely to the moment—for example, when you jump out of a plane with a parachute, when you race along a narrow ravine in a flapping wingsuit within a hair's breadth of cliffs rising up on either side, or when you climb up the outside of a skyscraper to the very top without a safety harness. But, then again, what's extreme about lying on the grass with seven dozing goslings?

I allow my thoughts to wander, look up at the sky and the edge of the wood, and then, out of the corner of my eye, I catch sight of a large pheasant stepping out of the undergrowth about a hundred yards away. It struts calmly out into the field, the tips of its feathers glinting in the morning sun. It looks totally at ease. Then it looks up, lifts one foot, and with a flap of its wings, disappears.

6

The Monster in the Barn

WAVING MY YELLOW JACKET MENACINGLY AROUND MY head with one hand and madly honking my horn with the other, I yell, "Come, come, come!" and gallop off like a man possessed, until we reach a couple of apple trees at the edge of the meadow. Seven alarmed goslings waddle in my wake.

We've escaped! Barely. We catch our breath under the trees, which, as the little geese are well aware, offer some protection from an aerial attack. They gather around my feet and cuddle up to each other, relieved.

I hold up my hand to shield my eyes from the sun and scan the meadow to see if anyone has dispatched an emergency vehicle from the psychiatric hospital to collect me, for anyone watching me and the geese just now would probably think that I'm completely out of my mind.

JUST A MOMENT EARLIER, EVERYTHING HAD LOOKED SO calm and peaceful. We'd been meandering through the large meadow close to the castle, which I've decided to call Castle Meadow. I'd sought out a lovely little spot and spread out my camping mat, delighted by the idea that this was the place where I would soon be teaching the geese to fly.

"So the runway will be about here," I explained to the goslings as I sat down. "Then it's up and away into the clouds. Isn't that great? Right? Won't it be awesome to finally be up there flying?"

I watched Nemo with pride as he gazed in fascination up into the sky. He clearly appeared to be sharing my enthusiasm.

"Way up there. All of us together. Weeweeweewee," I repeated. "That's the third dimension."

Then I looked up and got an awful shock. Two kites and a buzzard were circling above our heads in anticipation of a feast. And here I was, calmly parading around with seven twenty-ounce portions of prime-quality organic goose meat ready to eat and attractively laid out on a grassy salver. How could I be so negligent? This would never have happened to a real father goose.

Unlike me, the goslings were skeptical from the moment they spotted the large expanse of mown grass and had stayed unusually close to me, something I have to admit I hadn't noticed at first. A young goose doesn't stand a chance against a buzzard, and the more offspring a mother goose has, the more difficult it is for her to protect all her babies from raptor attacks. Once a buzzard has selected its prey, it dives down and kills the little one with its talons before the gosling even realizes what's happening to it. Then the buzzard enjoys its meal of goose meat in a secluded spot, often regurgitating it later for its own chicks, because buzzards are altricial and rely on their parents to feed them. That is the food chain as nature intended.

Luckily, we just managed to avoid this grim chain of events. The raptors obviously hadn't factored my loud shrieks and the wild swings of my yellow jacket into the equation. *Don't mess with Goose Michael!* When I finally peek out from under the apple tree, the kites and the buzzard have disappeared.

• • • • • • • • •

WE ORIGINALLY CAME TO CASTLE MEADOW BECAUSE I HAD a completely different monster in mind: my Atos—that is to say my ultralight—in which I soon hope to be flying along next to the geese. After all, flying and collecting measurements while flying are the reasons for all our efforts, and the geese need to get used to the plane as soon as possible. That's why we stationed it in an old barn next to Castle Meadow a few weeks ago. For the first few weeks, the barn will be our temporary hangar, one that we can easily reach by waddling—unlike the airfield, which is too far to get to on foot. The ultralight needs to become a completely normal part of our daily goose routine, so I want to spend as much time as possible in its vicinity.

Although the Atos is relatively light for a plane—just barely two hundred pounds with a twenty-seven PS motor (which is roughly equivalent to a twenty-seven horsepower motor)—its almost forty-foot wingspan must make it look like an enormous monster to the little geese. This sublime aircraft, however, delights me every time I see it. Who, back at the turn of the twenty-first century, would have thought that one day you would be able to fly using nothing but electricity? Without any smell, sound, or exhaust fumes? The battery is charged using hydroelectricity, and once it's charged, the ultralight can fly for about an hour.

The plan is first to get the geese used to the monster. Every day, I'll taxi the Atos a short distance out into the meadow while the geese walk alongside me. If that works well, I can take them to the real airfield. A side benefit is that once the plane is out of the barn, we'll no longer have to worry about raptor attacks. The mere sight of the ultralight should be enough to instill respect in the buzzards and kites that are so numerous here.

The thing I'm most excited about is how the geese will react to the sound of the propeller, which I kept playing to the goslings when they were still in their eggs. Did the imprinting work as well for the propeller as it did for the horn?

I slide open the heavy roller door, and there in front of me is a flying machine made out of carbon fiber and canvas. The geese sense that this high-tech structure doesn't belong in an old barn among dusty agricultural equipment, rakes, and brooms. In these surroundings, it looks surreal. A bit unnerved, they huddle behind me, as though they're afraid the thing with wings could attack them at any moment.

As I slowly push the plane out of the barn, however, their fear quickly dissipates. Perhaps at that moment it becomes clear to the geese how imperfect and clumsy the high-tech machine is in comparison with the wings of a mature goose. I think I detect disdain in their eyes as they give the plane the once over. It doesn't even seem to surprise them when I climb into the machine. The electric motor beeps.

I whisper, "Weeweeweewee. Stay calm, my little geese."

Then I carefully push the accelerator lever forward, and the propeller begins to turn with a loud rattle. The geese are spooked for a moment and jump back a couple of feet.

I honk my horn and keep calling, "Weeweeweewee. Come, come, come, geese!"

They come close again, despite the propeller being quite noisy as it whirls around right behind me. Nemo, Gloria, and the ever-bold Calimero even let the breeze from the propeller blow over their faces.

I increase the rate at which the propeller is rotating, and the noise becomes a roar. I can't hear my own voice, but the geese don't care at all. They stay sitting in the grass next to the plane, completely relaxed. Getting them used to the noise has worked extremely well. I'm really proud of my seven courageous little geese.

What was that?

The dude just dragged out something else he wants to show us.

I think he's nuts.

I know. But don't tell him.

Let's just pretend we're really interested.

It doesn't take long for them to subject the plane to a rigorous inspection to see how it holds up to being nibbled. I remain seated in the plane as I watch the little ones. Before the geese lose interest, I really want to try a sprint across the meadow with them, because when we start flying, they've got to be capable of getting up some speed. And so I get out, strut importantly into the middle of the makeshift runway.

"Ready? Pay attention," I shout.

Then I honk the horn loudly three times and take off on foot. The little geese run after me like world-class sprinters. It looks as though someone has released seven golden yellow tennis balls and they're spilling out across the meadow. We repeat the sprint a few times, and then I allow the geese to take a breather in the shade of the Atos's wing. After all, I don't want to ask too much of them. I push the plane back into the barn, and we set out for home.

THE NEXT DAY DOESN'T BEGIN UNTIL SIX. I RUB MY EYES in amazement and cast a questioning glance at the geese. Did the cuckoo die? Or did he just decide to be more reasonable? Either way, if this is the new normal, that's fine by me.

Today's the day we start training for takeoff. I want to try taxiing the Atos alongside the geese. This is a potentially risky maneuver, because unlike a real mother, like Karlsson-on-the-Roof—another character created by the author of Nils's adventures—I need a propeller on my back to achieve lift-off, and I have to make sure that the little geese stay away from it. Right now at least, the distance between the tips of the blades and the ground is far enough that they can't possibly reach the propeller.

We stroll up to Castle Meadow—not, of course, without making a short stop at our favorite field of dandelions on the way—and I push the Atos out of the barn. This time the geese are noticeably less apprehensive. They seem to recognize the plane.

My plans with the Atos will be asking a lot of the geese, and they need to learn how important it is for them to not let me out of their sight. Therefore, I purposely push the plane some distance away from them without calling or honking the horn.

After a while, it dawns on them that they're alone, and they start peeping for me anxiously. I find it very difficult not to respond to their calls right away, but I let them cry for about another fifteen seconds before I honk the horn and call, “Come, geese!” It warms my heart every time I watch the seven baby geese run to me right away with their stubby little wings outstretched, happy to be reunited with their papa.

I plan to position the geese in front of me and off to one side. Then, I’ll taxi gently forward, honking the horn loudly while calling “Come, come!” If one of the little geese falls behind or comes anywhere near the propeller, I’ll shut off the motor immediately.

The geese are grazing peacefully and don’t even notice as I climb up into the pilot’s seat. That’s fine by me, because it means I can devote myself completely to thinking about what I’m going to do next. What’s the best place to position the geese? How fast should I accelerate? I absolutely must not run over a single one of them.

I turn on the electrical system and the steering mechanism begins to beep. That tells me that the motor is engaged and ready to go.

“Come, geese!” I call.

I urge them into position diagonally in front of me, a few feet off to one side of the front wheel.

I open the throttle slowly. There’s a gentle humming noise, and the geese startle. By talking to them and honking the horn, I quickly manage to calm them down again. I increase the rpms and feel the propulsion from behind. The propeller pushes the Atos forward, and the plane and I begin to bounce very slowly across the meadow. Nemo and Calimero give me a questioning look.

"Come, geese!" I call. "Off we go! Off we go!"

As though it were the most natural thing in the world, they start walking calmly alongside the plane.

"Great," I call. "Just like that! Super! Come, geese!"

I'm so proud, I allow myself to get carried away and speed up a bit. The Atos suddenly lurches forward, and in a single bound, I've overtaken the geese. I turn and see them about a dozen feet behind me. Everything's just fine. The geese even seem to find it quite funny. Instinctively, they've avoided the propeller and the rear wheel, and they're still cheerfully walking along.

I increase the speed a bit more, and the geese run as fast as they can behind me and the plane. When I turn around again, I get a fright. One of the geese is lying on its back in the grass well behind the others, flailing its feet in the air.

I come to a complete stop and jump out of the plane. Who is it? Did one of them bump into the propeller after all? When I finally reach the goose, it takes me a minute to figure out what's happened. It's Nils, who's got himself stuck in a rut made by a tractor tire and is now lying on his back, unable to right himself on his own. So he just tripped while he was running along, and unluckily for him, fell down into the rut.

"Nils," I say relieved. "You just tripped. Thank goodness!"

I free him from his awkward position and call to the rest of the geese.

"You were great. That was amazing. In a couple of days everything will go swimmingly, but today it's time to go home."

• • • • • • • • •

"COME ON, GEESE. HURRY UP!"

We're getting close to the camping trailer. I can hardly wait, for the simple reason that I have a pressing need to relieve myself. As soon as we're inside, the little geese fall on their food and I rush to the bathroom. But no sooner have I sat down than I hear them crying pitifully outside, so I open the door and shout at them rather crossly.

"I'm right here! Oh, just come right on in, why don't you?"

Two seconds later, the seven scamps are gathered around my feet. They cuddle up under my dropped pants, and their sleepy whistling begins. For me this means half an hour trapped on the toilet unable to move. After fifteen minutes, I have pins and needles in my legs. But then things improve, because after twenty minutes, my legs are numb and I can't feel anything anymore.

Seven Differences between Geese and People

1. They don't need toilets.
2. They don't knock.
3. They don't have any difficulty sleeping.
4. They know what's good for them.

5. They don't use us for research.
6. A gander could never be a father to human children.
7. When they want to get closer, they just do it.

• • • • • • • • •

WHEN THE TROOPS HAVE FINALLY FINISHED THEIR BEAUTY sleep, they stand peeping expectantly at the bathroom door, as though saying: *What on earth have you been doing in here for such a long time?* Meanwhile, I have no idea how I'm going to get up now that I've lost my legs. I have to drag myself up with my arms, and then I hit my legs and stamp one, and then the other, for quite some time. The geese observe me, amused.

What's wrong with the dude this time? Is he drunk?

I think he's dancing.

But why?

I don't know if I want to know why.

• • • • • • • • •

BAM! AN EARSPLITTING CRACK JOLTS ME FROM MY SLEEP at three in the morning, and I soon realize that the thunderstorm wasn't a dream after all. A moment ago, I was watching the camping trailer from outside as it floundered, sails flailing, on a savage sea. The geese were frantically swimming after it, and there was an insistent honking coming from somewhere.

I rush outside. Rain is pelting down against the outside of the camping trailer. A flash of lightning lights up the scene for an instant, and I can see the awning flapping in the storm, looking as though it could fly away at any moment. I jump into my rain gear, grab a hammer, and drive the tent pegs securing the awning back into the ground with a few hefty blows.

Unfortunately, that doesn't help at all. The ground is too sodden to hold the pegs, and the fury of the storm rips them right out again. All I can do is cut the guy lines with my pocketknife and lash the awning to the camping trailer. That works quite well, until I realize that now I can't get back in, because I covered the trailer door when I tied the awning down. That's almost as stupid as sawing off the branch you're sitting on.

I grab my camping mat and run to my geese in the aviary. They look at me somewhat apologetically when I open the door, because all seven of them are huddled together in the driest corner of their quarters. But as soon as I lie down they come and cuddle up around my head. Together, we stick it out while outside the thunder roars, lightning flashes light up the sky, and hail hammers relentlessly on the roof. After half an hour, the storm passes.

I get up from my uncomfortable position, bid the geese a good night, and clear a way past the awning into the camping trailer. I collapse into bed, exhausted. I'm completely soaked but happy that my geese and I have survived the storm together.

7

Geese in the Van

"HOP TO IT! COME ON!"

"Go for it! Frieda, hop, hop! Calimero!"

I'm perched on a makeshift ramp at the back of a Volkswagen van, sweet-talking my little geese.

"You can do it! Up we come!"

"It's not so bad in here."

Despite my dulcet tones, the geese remain skeptical and examine the two rubber-cushioned planks suspiciously.

Hey Calimero, do you see that shit there? Is that some kind of waddling aid? What crazy idea has gotten into his head this time?

Just don't even think about it, Nemo. Just don't think about it.

Today's an important day. I want the geese to get into the van so that I can drive them to the airfield for the first time. The runway training in Castle Meadow is going extremely well. The little geese are running behind the plane at an impressive speed, madly beating their little wings as they go. But there's simply not enough room in the meadow for the first real practice sessions with the ultralight. If I take off there, even briefly, we'll all end up in the trees, and I want to spare us that traumatic experience.

The geese already know how to do an amazing number of things, but the straightforward task of hopping into a Volkswagen van isn't part of their genetic programming. That's why I've come up with the idea of this elevated waddle ramp. But it's easier said than done. The ramp mustn't be too steep, which means that the planks have to be quite long. And the longer the planks, the greater the danger the geese will jump off the side while they're waddling up it.

Behind the camping trailer, I found two planks covered with rubber matting, each about six feet long, and I'm sitting on them right now. Theoretically, the geese can now waddle comfortably up into the back of the van.

"Nemo! We're driving off in the vaaaan. To the airfield. Just walk right on iiin!"

I'm feeling rather silly, but then I produce my secret weapon: a dish of delicious gosling starter feed.

"There's food up heeere!" I coo.

As strange as they find the VW-monster, there comes a point at which the geese can no longer resist. Frieda is the first to dare a couple of goose steps up to the van, while I waddle slowly backward into the back of the van, scattering feed as I go.

"Yes, yes, yes. Calimero! That's great! Just keep your mind on the food! There's more up here!"

"Only a couple of feet to go! You're almost there!"

The geese peck and waddle, and I'm just about to start breathing again when Frieda gives a loud neurotic shriek, jumps off the ramp in a single bound, and hides under the bus.

"Frieda!"

My other geese, naturally, let her anxiety get to them. It occurs to all of them all of a sudden that normal geese don't climb into vw vans and that such behavior is totally unnatural and unusual. They scatter, honking loudly.

"Man, Frieda!" I moan. "Can't you ever relax? Do you always have to get the others worked up? Goose fat here we come!"

The geese withdraw into the shadows under the camping trailer. Now I'll never get them in the van, and so I give up for the day. Earlier, the behavior of the geese would have annoyed me, and I would probably have been quite frustrated with them for the rest of the day. But now, I think: *Things aren't so bad. We'll just hang out on the sunny deck instead.*

My newly acquired serenity feels great. I sit down on the bench under the awning, feeling a bit tired, and consider how best to deal with Frieda, the problem goose. She's slowly beginning to disrupt the group, and I'm worried that her behavior will undermine my authority. What would a real mama goose do in this situation? Should I give Frieda even more cuddles or would it be better to crack down on her? Or is she simply going through a completely normal preteen phase?

I lie down in the meadow, rest my chin on my hands, and observe the gaggle of geese under the camping trailer. Frieda's preening and doesn't look at me. Her movements seem somewhat staged and defiant, like a person who starts dramatically cleaning a room after an argument. Or am I just imagining things? That's way too much information to be getting from a goose.

• • • • • • • • •

AFTER FIVE WEEKS, THE GEESE ARE QUITE BIG, THOUGH A few of them still have a somewhat bizarre hairstyle, with puffs of yellow down—remnants of their initial growth of feathers—still fluffed out on their necks. When I heaved Nemo up onto the scale yesterday, I could hardly believe how heavy he is. He's already broken the 5-pound mark. A regular goose dumpling. In only five weeks, he has increased his birth weight of 4.2 ounces by a factor of twenty. A human baby growing at the same rate would weigh nearly 155 pounds at five weeks.

Granted, the protein content of the nutrient-rich gosling starter feed that I give the geese every day is a hearty 21 percent, which is optimal for the strength and development of the goslings. But the feed is not the only reason they're putting on so much weight. Geese simply grow astoundingly fast.

The birds' initial downy fluff has already been replaced by proper feathers. Goose feathers are marvels of engineering. They are airtight and watertight, insulating, and wonderfully light. And they are formed without relying on solvents or synthetics, from nothing but water and protein. Just think of how much it effort it would take to create an artificial goose feather.

I call to Calimero, who's the one peeping out from under the camping trailer right now.

"You're big enough to know better, you lousy good-for-nothing goose."

In response, he briefly waggles his rear end and shoots a quarter of an ounce of the finest fermented greens mixed with digested grains and a bit of mealworm chitin at my feet. I'm using the word "shoots" intentionally. The force with which the geese can discharge such loads has to be seen to be believed. Even though it's lovely to have seven geese cuddle up to you, you do look shitty afterwards. And geese discharge their droppings every eight to ten minutes. That makes for a whole pile of doo-doo in a day.

The discharges of my geese play a central role in my hermit-like existence. Over time, I've become inured to them, and what I think about most now is the unbelievable color palette of goose droppings. Even goose shit is one of nature's wonders. The color isn't simply a monotone green. The hues range from dark brown (indicating the consumption of an ample amount of grain) to dark green (lots of greens). And let's not forget the ever-present white parts, which are actually uric acid and not, strictly speaking, fecal matter at all. In a goose, the relatively short urethra empties into the gut at the cloaca, and urine is excreted from this opening along with the feces. And that's why goose droppings are flecked with white.

A while ago, I watched a TV program about two young Danish designers who made furniture from seaweed. They boiled the seaweed, mixed it with pulped cellulose, rolled it out, and, using a homemade oven heated to 104 degrees Fahrenheit, dried the mixture out to fashion a chair back. The designers were proud of the marbled effect created by the different shades of brown and green in their products.

I observe Calimero, who is pointing his rear end in my direction once again, and think: *That could work with the geese.* Furniture from goose droppings. And then the clincher. Just as beekeepers have their bees fly over a field of rapeseed so that they can offer rapeseed honey, I could—according to the wishes of individual clients—allow the geese to digest only certain greens, and then I could offer chairs made from dandelion or clover, pulped and organically processed in goose gut.

And just as I have become inured to goose droppings, I also no longer take much notice of the multitude of blackflies, plant-loving bugs, and gnats that are constantly milling about in the vicinity of the camping trailer around me and my family of geese. No doubt I'm gradually mutating into some sort of waterfowl. Rather than resorting to a chemical deterrent, I seriously consider covering myself in mud. There's enough of it around, and dried mud functions as a natural protective coating. Or I could rub myself with aromatic herbs and plants. But as I'm botanically illiterate, I would probably confuse lemon balm with poison sumac. At least I still take a shower every day and don't try to clean myself with my bill like a goose. As inflexible as I am, I wouldn't be able to reach certain body parts that absolutely should be cleaned every day.

• • • • • • • • •

THE NEXT MORNING, MY HERMIT-LIKE EXISTENCE IS INTERrupted. I'm getting help from Laura, a young assistant at the Institute. When we take off from the airfield for the first time, I'm going to need someone else to gather the geese near the plane.

I don't expect Laura at the camping trailer until ten. Until then, there's plenty of time for us early risers to take a short turn around a neighboring marsh. The geese love it when I lie in the swampy reed beds on my camping mat while they dabble contentedly in the soft ground and find fresh greens to nibble on. After about twenty minutes, one little goose after the other comes to cuddle with me on the mat. For a while now, Gloria, the oldest, has reserved the prime spot on my lap for herself. The rest of the geese make themselves comfortable next to me or on top of me.

It takes no more than five minutes for them all to be whistling sleepily through their bills, and then I can't move. Luckily, an arm that's falling asleep soon becomes an arm that's completely numb. And so I lie there, one armed, feeling Nils's breath on my earlobe. He loves to cuddle his head into my neck under my hair—which really tickles. I endure this along with the loss of one of my arms.

Pinned by the geese, I lie with the morning sun shining in my face. Around me in the muck, insects are buzzing and chirping, every one of the tiny creatures a work of art. The air smells of grass and meadow flowers inviting bees to visit them. Striated cirrus clouds move slowly across the sky way up above our heads.

• • • • • • • • •

I WAKE UP, BECAUSE "RAMBO" IS ENGAGED IN BATTLE with my earlobe. Calimero is determined to pull out my earring and goes about his business in a fairly brutal manner. I turn over carefully, push him off, and look at my watch. Shit! It's already ten thirty. How could I have slept for so long? I normally don't manage to relax so completely out in the fields, because I worry so much about dogs and raptors having designs on my little geese.

I have to hurry. With one arm still hanging uselessly at my right side, we march in single file back to the camping trailer, where Laura greets us cheerily. Laura is a biologist with short hair who always has an ironic gleam in her eye. She has worked with blackbirds, carrier pigeons, and chickadees, and she seems completely unfazed by my shitty appearance. She's sitting on the deck in front of the camping trailer with her legs stretched out in front of her.

"Ah, here come the eight geese," she teases as she greets us.

"We fell asleep," I explain.

"You and the geese, both?"

Luckily, the geese are very accommodating and not overly shy. They hang behind me for a while but go to Laura when she produces the basket of dandelion leaves she has brought with her. Even if she wanted to, after five weeks Laura couldn't step in as a stepmum in this blended goose family. The geese would certainly get used to her, but they would never accept her as a mother in the same way they've accepted me as their father. It's too late for that. The time for imprinting is long past.

Despite this, the geese are curious, and it doesn't take long for them to start exploring Laura's shoelaces with their bills. Playing with shoelaces has recently become one of their favorite activities. They just can't get enough of it. You could say they're addicted to shoelaces.

As soon as I put on shoes with laces, the little ones fight to see who'll be the first to nibble and pull on them. Nemo is usually the fastest, and the two of us play "Tying the Shoelaces." I tie the laces and he nibbles them loose with his bill. Over time, he's become a world-class untier of laces. It takes him no more than a couple of seconds, and he never tires of the game.

In contrast, it would never occur to Paula to fight with Gloria and Calimero over the shoelaces in my right shoe. She overcomes the "disadvantage" of her low rank in her own way—with cunning. She simply toddles around to my left shoe, which the others have completely overlooked in their wrangling over the right one.

SO THAT WE DIDN'T WASTE MORE HOURS DOING A BALANCING act on the ramp and enticing the geese into the van, Laura has brought a large dog crate along with her. Once the geese are inside, all we have to do is lift the travel crate into the van. This has the added advantage that the geese will be confined during the drive and not flapping about in the back of the van.

I open the doors to the crate and place a plate of delicious grain exactly in the middle. At first, the geese stand around outside the crate looking surprised, nibbling at it from all sides. Then the hierarchy in the group comes to the fore. Nemo is the first for whom greed wins out over suspicion. For a while, he has the bowl of food all to himself. Then Calimero and Gloria join him. The rest of the gang shift their weight from one foot to the other outside the crate. Eventually, Paula summons up courage, and Nils, Maddin, and Frieda follow her in. The seven of them tuck into the grain while I shut the doors to the crate and stay sitting beside them. That was definitely easier than using the ramp.

Luckily, there aren't any more major problems getting the geese to accept their mode of transport. However, I must admit that we've grossly underestimated the weight of the crate once the geese are inside, and we have to drive right up to the camping trailer so that we can heave the crate directly into the van.

"You can ride in the back and hold the geese by the hand," laughs Laura, as she climbs up front. "Like a paramedic in an ambulance."

The geese take no notice of her and continue to happily nibble away. But as soon as Laura starts to drive, they sense that their goose might be cooked, and they begin to protest. Driving is something completely different from swimming or waddling. It's a mode of locomotion that is completely unfamiliar to them. Although it has to be said that not all of the geese are protesting, and the one who is protesting the most is, of course, Frieda.

"Eeek, eeek, eeek!" she shrieks.

"Frieda! It's only a van. Vroom, vroom," I explain. "We're driving to the airfield."

Even though I add arm movements to imitate flying, Frieda's anxious shrieks resonate more with the other geese than my calming words.

"Eeek, eeek, eeek!" all seven are soon protesting.

"Weeweeweewee," I reply.

We drive through the countryside like a beeping clown car. All the van windows are down because of the heat, and people on the roadside stop to gawk in amazement.

They probably think I'm out of my mind because, from the outside, you can't see the geese. All people can see is a guy making strange noises and calling out names.

"Frieda! Be quiet! We're almost there!"

"Calimero! Weeweeweewee!"

"Eeek! Eeek!"

"Everything okay back there?"

We only have to drive two and a half miles, but it feels like we'll never get there.

I think it's great when other people participate in our adventure and want to watch, but when we reach the airfield, I'm glad that today there's not a soul in sight. We could do with as few sources of distraction and diversion as possible on our first day of airfield training. In addition, it's already very warm at the airfield, unfortunately, and I have my doubts about being able to motivate the geese. When it's really hot, they prefer to lie around in the shade near water. I don't blame them. Who would want to chase after some lazy old fart for no discernible reason when it's eighty-six degrees Fahrenheit and you're wearing a down coat?

"What are we going to do," Laura asks, "if the birds are so out of sorts after the drive that they simply run away as soon as we let them out of the crate?"

"Good question," I reply. "But they won't."

"Even so. What could we do if they did?"

"Mmmm."

As it so happens, I've asked myself this same question many times.

"There's nothing we can do," I reply eventually. "But if I want to get up into the air with them, we have no choice."

Laura doesn't seem completely convinced.

"Let's go and get the big bird first," she suggests.

We push the Atos out. Assuming we're ready for the next step, someone from the Institute has moved it from the barn to a real hangar. Then I open the doors to the crate. Stepping carefully, the curious geese enter their new terrain. The thing they like best is the enormous expanse of fresh grass still wet with dew. All of them seem optimistic about how this is going to turn out—all except one. Frieda immediately withdraws under the van and sulks.

"Jeez, Frieda. Not again!"

"She's got you under her thumb," Laura remarks, and she's right about that.

"I think there's going to be roast goose with red cabbage for dinner tonight," I retort.

But of course I don't really mean that. After living with the geese, I no longer eat fowl of any kind. When a colleague recently brought me nasi goreng from a local Indonesian restaurant, I had to pick out the pieces of chicken. There was no way I could choke them down. I don't know exactly why, but the idea of eating chicken made me feel nauseous.

Quite apart from the fact that I have no intention of eating Frieda, I have to say that I find nonconformist behavior rather commendable. I'm not running a goose boot camp after all. As far as I'm concerned, Frieda can stay sitting under the van exercising her right to protest all day long. I'd even bring her a banner and a length of chain in case she wants to chain herself to something. But here's the problem. If Frieda the revolutionary won't participate in runway training, then the others won't either, and that puts our whole schedule in jeopardy. Her mutinous gabbling is spoiling the fun for everyone.

When Frieda finally crawls out from under the van, we're immediately faced with another problem. It's really windy. I'm sitting in the plane and I have my hands full trying to keep the wings off the ground, not letting the geese out of my sight, and finding the right moment to start.

The Atos's wide wingspan presents the wind with a large surface area to buffet, so I'm having difficulty keeping the ultralight balanced. It takes all my strength to stabilize the wings in the repeated gusts so that they remain somewhat horizontal and the wingtips don't hit the ground. It's particularly exhausting when the plane is moving slowly and the air is not flowing aerodynamically over the wings. At higher speeds, the rudder, which also acts as a spoiler, helps stabilize the wings, and I don't have to expend as much energy to keep them balanced.

While I wrestle with the plane, the geese are dashing all around me. Laura is driving them to the right-hand side of the plane, her arms extended with broomsticks. The wind is giving me a lot to deal with, but the geese are doing really well. The runway training has paid off, because the geese have already developed a routine to follow. Once the geese are in position, and Laura has removed herself and her long broomstick arms out of the danger zone around the propeller, I yell somewhat hysterically.

"Off we go, geese! Propeller clear!"

Then I honk my horn frantically, like an overexcited taxi driver, and accelerate. The geese immediately start moving, as though they're being remotely controlled. It's working. Together we dash up and down the runway. I'm proud of my little ones. They avoid the propeller all on their own and don't come too close to the wheels.

As they run, they flap their little wings like mad. Perhaps it bugs them that their wings can't produce enough lift yet. They can't take off no matter how hard they flap their wings. Someday I want to conquer the sky with the geese, but for now, what we all need is patience.

• • • • • • • • •

LATER, WE ALL LIE TOGETHER IN THE SHADE OF THE ATOS'S wings and cool our jets with delicious spring water, grain, and chocolate bars. The temperature has already reached eighty-six degrees, so it's better that we break off for today. There's no way I want to risk a meltdown, and the birds need to leave the airfield with a positive feeling after this first visit.

"Geese, do you want to go to the lake?" I ask.

To be quite honest, I, too, am yearning for a quick restorative dip in the nearby swimming lake—to say nothing of the fact that there's a kiosk next to the sunbathing area where all kinds of exquisite treats are available, including currywurst with a side of fries. First, I just have to honk the team of geese to order.

"If we want to go to the lake, then you have to get back into the crate," I explain.

Perhaps the geese don't understand what a "lake" is, but they're completely unconvinced by my argument. Laura and I join forces and attempt to drive the geese into the crate using broomsticks, but we fail miserably. There's only one thing for it, we must try to catch the little hellions by hand.

If you've ever tried to grab a chicken, you can imagine that this isn't going to be easy. To make matters worse, the geese aren't at all afraid of us; they just don't want to be caught. When I reach out to grab Frieda, she just goes, "Eeek, eeek, eeek" and runs away. It's only once we've trapped the geese all together in the confines of the carport that it dawns on the unruly band that there's no escape, and they acquiesce to the travel crate and my shuttle service. When we arrive, I collapse onto the grass covered in sweat and completely exhausted. I'm famished.

8

Through the Eyes of Children

FORTY SUNBATHERS ARE RELAXING ON THE GRASSY AREA BY the lake, talking about everyday people things, reading, making phone calls, and tapping into their cell phones. Suddenly, a wiry woman, a scruffy guy with unwashed hair and goose shit on his jacket, and seven greylag goslings enter stage right. We might as well have ridden in on an elephant.

Greylag geese are common around here, but they normally don't dare approach the sunbathing lawn, and they're not normally waddling obediently along behind a person. People are shaking their heads and pointing at us, but their astonishment is short-lived. The little geese soon win everyone over, and magically the sunbathers start to smile. Children and adults approach to within a few yards of us to observe the unusually trusting geese. We're surrounded by merriment and curiosity. The lake is suddenly overflowing with a strange and unusual sense of intimacy between people and geese.

The geese, however, are hugely unimpressed by the attention and the looks they're getting. They couldn't care less if people are staring at them. Because I'm their father goose, they are, of course, far less skittish around people than grown-up wild geese would be. All they want to do is get into the water right away. But even a father goose can't always put the needs of his children first. I'm ravenous, and I can already smell those currywurst. And so I march directly up to the kiosk with the geese dragging rather reluctantly along behind me.

Didn't I tell you? Do you smell that? Now we're getting to the bullshit that dude's been planning all along.

You could well be right, Frieda, but I haven't got any choice. He's got the horn!

And if the butcher had a horn like that, Maddin? Would you still be waddling along behind him?

Ten minutes later, I'm shoving the first piece of currywurst into my mouth, and I'm transported with delight. Delicious!

"Why hasn't a fox ever tried to get a few geese to imprint on it?" Laura wonders aloud.

But I'm too busy keeping a close eye on Calimero to follow her line of thought. Naturally, little Rambo immediately has to check out this curious, warm, elongated, and somewhat hostile-looking object in a cardboard tray that is sending his father into such rapture.

"Imagine," Laura continues, "that a fox comes across a goose egg just as it's hatching. Then the gosling thinks for a few seconds, 'Mama, just look at you!' and then it's eaten by what it thinks is its mom."

I frown.

"Ummm. I don't think that happens all that often."

"Even so, I think that's sad."

"No, Calimero! Get your bill out of there! My food!" I snarl at him.

But he has already stuck his bill into the ketchup-curry sauce. The onlooking sunbathers find what happens next particularly amusing. Flapping his wings and honking, Calimero hops across the lawn like a rabbit and tries to wipe his bill off on the grass. I jump up and run to him. He's so surprised by the odd taste that he lets me pick him up without making a fuss. Normally, he wouldn't allow me to do this—unlike cuddly Paula, for example. I rinse off his bill in the lake. Unfortunately, things that delight people—like currywurst, for example—aren't always good for geese.

After I've managed to rinse off the biggest blobs of ketchup, I set Calimero down at the edge of the lake. Immediately, he begins an orgy of cleaning, slapping the water with his wings until there's spray everywhere. The sunbathers sit on their towels grinning and watching the show. Finally, the mother of a boy of about five screws up her courage and starts up a conversation with us.

"Why are the ducks following you everywhere you go?" she asks, looking at us in amazement. "I've never seen anything like it."

"I've imprinted the geese on me," I explain.

"How does that work?"

"It's actually not that difficult. I just have to be the first thing the geese see and smell right after they hatch. It works even better if they can hear my voice before they hatch, while they're still in the egg. And then I need to give them lots of care and attention."

"That's all?"

"Well, the geese and I live together in a camping trailer. At first, I couldn't leave them on their own. That's quite a commitment."

"So it's like looking after a baby?"

"Yes, except it's over sooner. Much sooner."

The woman looks at the geese for a while.

"Can Enrico stroke the geese?"

She means her son, who's been standing somewhat shyly behind her legs the whole time—just as the little geese used to do with me.

"Of course," I say as I look around.

What I need now is a model goose. Frieda, it goes without saying, is not an option. Calimero, who's still struggling with the aftertaste of ketchup, is also out of the running. I walk over to my loveable, obedient Paula and take her in my arms. I can rely on her to not cause any trouble. She allows me to pick her up without any problem and immediately starts tugging affectionately and gently at my hair.

"This is Paula," I say to Enrico as I lean over to bring the goose down to his level. "She's lovely and soft."

Enrico carefully extends his small hand toward the goose and touches her feathers. There's hardly anything more beautiful than watching a child do something for the first time. Enrico is curious rather than fearful. He's also very careful. As soon as he feels the texture and warmth of the goose feathers, his face lights up with wonder and surprise—an expression you seldom see in adults. For me, living with the geese means more opportunity to look like this. More chances to remind myself of the beautiful and simple things that people usually miss.

I take Enrico's hand and guide it up under one of Paula's wings. It's lovely and warm here, and particularly soft and cozy, because of the many tiny downy feathers. Enrico beams.

"That was very brave of you," I say as the boy trots back to his mother.

I feel very happy. You don't need apps on a smartphone or a tablet to experience moments like this.

In the meantime, Calimero has calmed down. He gives the corpus delicti wursti a wide berth and walks up to the other little geese, who are lying lazily on the camping mat or picking at the grass. Paula and I lie down beside them, and soon I hear the familiar sleepy whistling of the geese.

"I'll be whistling like that soon," says Laura, "I'm that tired."

She lies down on the grass off to one side, so as to be outside the direct shit-firing line of the little geese. The birds gather around me and we doze off for a while. Only

Maddin is up and about, and he investigates the silvery object lying next to my head that sometimes calls the geese to Papa. He knocks his bill against it and nibbles on it. He moves on to investigate the black rubber bulb and soon elicits a tired but immediately recognizable honk. Maddin hops like a bunny to one side and stands there flabbergasted, his eyes popping out of his head. I can imagine what he's probably thinking.

He's been messing with us all this time. He can't even honk on his own!

Seven Things That Fascinate the Geese

1. Anything they can nibble and pull on
2. Shoelaces, especially red ones
3. Splashing and swimming
4. Grain
5. Grass, weeds, and dandelions
6. Shade
7. Papa's whereabouts

• • • • • • • • •

TWENTY MINUTES LATER, FRIEDA STANDS UP AND LEAVES A single, impressively large calling card on the camping mat to let us all know that—as far as she's concerned—it's time to get back to business.

The sunbathers watch us as we go down to the water together after our little nap. Nemo, Calimero, Gloria,

Frieda, Maddin, Nils, and Paula all follow me, one after the other, into the water and immediately begin splashing around. We swim together a short distance away from the sunbathing area to the shore nearby. Then I hear a loud jabbering and honking. This is exactly what I've been dreading for weeks. A real, bona fide biological family of wild geese is coming across the water toward us.

I suddenly feel like an awkward divorced stepdad whose adoptive children have unexpectedly been invited over by a picture-perfect two-parent family. I hope my children won't be jealous. I hope they won't realize that I'm not a real papa goose after all. I hope they won't run away. I hope they know that I love them ever so much. As the real mama goose swims by barely six feet away with her four little geese, I feel like a pitiful fraud.

"Hi!" I call. "Weeweeweewee," like one stressed mother goose might to another.

"Do your little ones get on your nerves, too, sometimes?" I want to ask. "Are they all different, as well?" The mama greylag goose looks a bit startled, but she keeps on swimming as though nothing strange just happened.

I don't want to risk any more of this, however. I call, "Come, geese, come" and swim freestyle as quickly as I can to shore so that at least I have my horn to call the geese away from the real mother goose. But none of this is necessary. My dear faithful goslings swim obediently after me and show no signs of doubting my paternity. I'm moved.

"Did you really think they would just up and leave you?" Laura laughs.

"Um, well, perhaps a bit," I reply, looking up at the sky.

The weather has taken a turn for the worse. The horizon's looking dark, and gusts are beginning to sweep over the lake behind us. We pack up hurriedly and march back to the parking lot. A thunderstorm is on its way, and I have no idea how the geese will handle thunder and lightning out in the open, so I really want to be back home with them in the aviary before the storm hits.

But how am I supposed to get Frieda back into the travel crate without the carport or some other physical barrier? I have absolutely no desire to play tag in a downpour with thunder and lightning going on around us. I'll have to resort to heavy-handed methods: I'll have to surprise her. I fall back a bit until I'm level with her. Then I grab her under her stomach with a hearty "So, stop right there!" and pick her up. She has no idea what's going on, and of course, she protests as loudly as she can. Clearly, this method isn't exactly helpful given Frieda's already difficult behavior, but I simply have no other choice. All it takes to convince the other six little geese is grain.

No sooner do we arrive home at the aviary than the skies crack mightily. Saint Peter is opening the floodgates. We seek shelter in the aviary, and Laura and I pass the time speculating what sex the geese really are. After all, you still can't tell by looking at them, so determining their sex is basically like reading tea leaves.

Some people swear that you can easily tell their sex by the shape of their feet. Others swear that geese and ganders waddle in distinctively different ways. Still others believe that the color of their feathers is proof positive of their sex. Meanwhile, Laura has had time to experience their personalities, and she now happily draws her own conclusions based on what she has observed.

"So Calimero is absolutely a boy," she says. "I'm sure of that."

"And I believe that Frieda is actually a rebel who wears pants, not skirts."

"Perhaps Frieda is just stubborn and excitable. That fits with her being a girl, too."

"There's no doubt about Paula. She's so timid and affectionate."

"But women aren't like that at all!" counters Laura.

Luckily, the horrendous storm is over in twenty minutes, and the sun is soon shining in the heavens once again. Laura bids farewell to each goose individually and calls to me as she leaves, "Bye, Goose Michael! See you soon!"

• • • • • • • • •

LAURA'S PRESENCE SEEMS TO HAVE DISRUPTED MY ENJOYment of my solitary existence. When I wish my beloved geese good night and shut up the aviary at about half past eight, I'm overcome by a need for the company of other people. I decide to leave the little geese alone for the very first time.

I ruin myself for goose company in my choice of evening wear—that is to say, I put on clothes with no goose shit on them at all—and I drive to Radolfzell, where I order a civilized glass of wheat beer right on the shores of Lake Constance. The café is fairly full.

At first, I enjoy being around all these people. But then I feel my stress levels rising. The people seem so frenzied and fidgety. And even though the atmosphere here on the lakeshore is so wonderfully beautiful, so satisfying, and so peaceful, their faces are devoid of emotion. Something important is lacking: the light I had seen earlier that day in Enrico's eyes. My time with the geese has affected the way I perceive my surroundings. My eyes have been opened. But what exactly do I see differently than before?

My thoughts turn to the chain link fence that borders the area around the camping trailer. A little while ago, I was sitting among the geese and glanced over to the fence, but all I saw were poppies in full bloom reaching for the sky. The scarlet flowers on their delicate dark green stems looked just wonderful. I wasn't even aware of the fence.

How often in our stressful, fast-moving daily lives do we focus on what is wrong, on the negative? Granted, there might be people who find green chain link fences more attractive than red poppies, but that's not the point. As *Homo sapiens*, we are so proud of our intellect that we sometimes forget that it also limits us. We consciously assess and judge everything. In the past, when I looked at a flower, I was often convinced I already knew what it smelled like, even though I hadn't gone over and taken in its scent. And so my intellect has killed off a wealth of experiences I might have had exploring the natural world. What would it be like to contemplate things more often from the nonjudgmental, instinctive perspective of animals? How vibrant and beautiful the world must look to a goose.

I look out over Lake Constance and realize that we can each decide for ourselves where we want to direct our attention. How would it be if we were to concentrate solely on the beautiful for once? For instance, there are all the different shades of green in the leaves on the trees. There are those wondrous goose wings that grow all by themselves. There's the feeling of peace that comes from being with the geese. There's the genuine need animals have for closeness.

And as I sit here and really notice how amazingly different the shades of green are in the leaves and grasses, I'm deeply moved and think how everything is one: the natural world in which we live, me, the geese, our different perspectives, and of course, the infinite variety of green in goose shit.

9

Geese and Ganders

"HANG ON TIGHT! THIS IS NO TIME TO BE TENTATIVE!" I shout as Nemo flaps his wings and tries to free himself from Laura's arms.

"Wow, he's strong," she gasps as she tightens her grip on Nemo.

I carefully insert the small thin needle, which is actually designed for insulin injections, at a slight angle into the protruding vein. It's relatively easy to draw the amount of blood I need, because a goose's veins are almost as large and thick as a person's. But Nemo is still relieved when he gets the procedure over with and can waddle away honking about how he's been tricked.

I've decided to clear up the question of the birds' sex once and for all. Even now, I could certainly try to find out what sex they are by turning their cloacas inside out, but that would be painful for the birds, and because of my inexperience sexing greylag geese, I still wouldn't be absolutely sure. So that's why I'm drawing a small amount of blood from each goose's thigh and sending it to a scientist friend in Heidelberg for analysis.

I'm taking this opportunity to weigh the birds, as well. Nemo is, once again, the heaviest. He weighs in at an impressive seven pounds and eleven and a half ounces. That's one pound and one and a half ounces more than the other geese on average. Nils, in contrast, weighs the least. That's nothing new, either. From the beginning he's been particularly small and delicate, which could be because he was the last to hatch.

Weighing and drawing blood go without a hitch with all the geese. Except one. What did I expect? Frieda, the drama queen, makes a great song and dance about it, of course. She acts as if, instead of administering a small prick, I'm going to throw her into a pressure cooker onto a bed of onions and sliced carrots. Talking to her lovingly has no effect whatsoever. It's only when Laura pins the flapping bird down with both hands that, after a couple of failed attempts, we manage to steal the required drops of blood.

"Frieeeeda! Keep calm. Everything's going to be fine," I whisper. "It'll soon be over."

"Gaagaagaa," squawks Frieda.

I might as well be talking to the birch tree outside. When we're done with her, we give her a heap of praise and a handful of grain as a reward. Although she continues to snort her displeasure, she does seem to enjoy the grain.

AFTER EIGHT WEEKS, THE BIRDS' FLIGHT FEATHERS ARE just about ready. It won't be long before the birds make their first forays off the ground. I can already see how they're strengthening their flight muscles daily by stretching their heads to the sky and madly beating their wings. Unfortunately, they're crestfallen every time they realize that however beautiful their wings are, they can't yet carry them up into the air.

"Micha?" I hear a voice call from outside the enclosure.

I jump up to go to the gate. For a moment, I'd almost forgotten that I'm surrounded by geese. As father goose, I can't just leap up and walk away. If I do, I immediately elicit a startle response from the alarmed geese. And that's why the whole rascally band comes along as a welcoming committee. Our visitor, Tom, is visibly amused as the goose family greets him.

Tom looks a bit like Crocodile Dundee, though he's actually a falconer from Bavaria who's become a friend. He's an expert in fitting little backpack cameras to eagles and other birds without adversely affecting the birds too much.

Last year, Tom fitted a high-res minicam onto the backs of two golden eagles—Arjas and Torron. He also developed a sophisticated carbon-fiber harness that meant the camera could be attached in such a way that the angle of the camera could be adjusted. And the very best part is that the harness fits so perfectly that it doesn't bother Arjas at all when he flies. The video from this camera is unbelievable: an experience from the perspective of a golden eagle flying high up over the Dolomites. When you watch it, you feel as if you're right there, flying through the air on the back of an eagle. And that's exactly what we want to try with a goose.

Tom's here to take a look at the geese and judge from the development of their feathers when they'll be able to fly. I've already noticed that when the geese flap their wings, they leave the ground briefly, but I'd like to get a better idea from Tom how long it will be before they can really take off.

Tom wears climbing boots with extraordinarily beautiful laces, and the geese naturally make a beeline for them. It's only after I've freed him from the clutches of the fowl mob that we can sit down over a cup of coffee and discuss our plans for the day. First, we want to try the carbon-fiber harness on Nemo's broad back.

Without further delay, I arm myself with treats and hold Nemo on my lap. He's eating the grain so greedily from my hand that he doesn't even notice when Tom puts the harness on his back. It's only when we do up the Velcro fasteners that he realizes there's something different about his rear end. But this doesn't make Nemo flap wildly. Instead he does something much more sensible: he begins to examine the high-tech apparatus on his back with his bill.

"Super, Nemo! All good! You're doing great!" I whisper reassuringly into his goose ear while giving him lots of strokes on his long neck. Then I extend his wings to show Tom how developed his flight feathers are. Tom gives his expert opinion.

"Any time now!"

"Really?"

I'm surprised and feel joyful anticipation on the one hand and on the other a slight pang of regret that the birds will soon be all grown up and fully fledged.

I take Nemo and his backpack carefully off my lap and set him down on the grass. His mood changes abruptly. He hops around the enclosure like a wild thing and tries with all his might to rid himself of the contraption on his back. He's probably just now noticed the slight pull on his thigh and realized that the harness doesn't have anything to do with me but is really attached to his body.

"That's completely normal," says Tom. "Just give him a few minutes."

"Poor thing," comments Laura.

I watch Nemo with a guilty conscience. On the one hand, I've raised him and I don't want to do anything that upsets him. On the other hand, raising him always was part of an experiment, and it goes with the territory that animals in experiments are subjected to stressful situations.

After barely five minutes, Nemo settles down on the grass, though he gives me a reproachful glance as he does so. We leave the apparatus on his back for another ten minutes and then I take if off—carefully—while rewarding him with a generous handful of grain.

TWO HOURS LATER, AFTER TAKING THE GEESE FOR A relaxing dip in the stream, Tom and I load them into the van and set off for the airfield. The geese know the ropes by now, so they're no longer scared. On the contrary, they keep up a volley of earsplitting honks of excitement from the back of the van. We let the geese out of the crate as soon as we arrive at the airfield. They immediately set to neatly mowing the lawn, as they always do.

Training for the geese has already settled into a routine. When I sit in the plane, ready to go, and honk the horn, they jump up, rush toward me like mad things, and, without any prompting, get into just the right position on the right-hand side under the wing.

Today, when I push the accelerator lever to the max, I'm amazed that for the very first time the geese aren't just running along behind me. They're so excited that they even run slightly ahead of me for a while. They're running faster than the Atos. After about five hundred feet, I ease off on the accelerator, turn the plane around, and repeat the process. This time, I aim directly for Tom.

Suddenly, I see him waving his arms above his head with delight as he begins to cheer. He's jumping up and down on the grass. What I'm doing with the plane and the geese today is great, I think, but is it really sooo amazingly special? I shut off the motor, take off my helmet, and give him a questioning look.

"Didn't you see that?" he splutters. "The goose with the orange leg band just flew for a dozen feet or more!"

"Really?"

"Yes! Absolutely! It was only a foot off the ground for about fifteen feet, but the goose was definitely in the air."

I turn to Calimero and look at him, thrilled.

"Woohoo. That was awesome! Weeweeweewee! I'm proud of you, Calimero! You're amazing!"

I flop down on my back in the grass and look up at the sky. The time has finally come. I'm flooded with a feeling of joy that takes me by surprise. For a long moment, I just lie there and stare at Calimero and the rest of the geese, as though they have some kind of a hold over me. And they really do. The sense of being grounded that emanates from them is nothing less than a gift. Calimero, of course, doesn't understand what all the fuss is about.

Chill, old man. That was nothing. One of the easiest things I've ever done.

I never thought Calimero would be the first to take off. I would have put my money on Gloria. But perhaps I underestimated Calimero when I cast him as Rambo. He's more like an elite goose SEAL.

We have a barbecue back at the camping trailer to celebrate. Not with Calimero, of course, but for him. As we celebrate, Frieda disappears under the camping trailer again to wallow in her neuroses.

"Frieda, dear, you stubborn goose. Come out and hang with us for a while. Come, Frieda. Come ooon," I call to her, to no avail.

Like a Greenpeace activist chained to a nuclear cooling tower, she disdainfully ignores my blandishments and refuses to leave her post.

"What can I say? A real Che Guevara of geese!"

"More like an ornery goose," Laura remarks.

At least the cobs of corn and the grilled vegetables taste delicious. Thanks to my isolated existence in the camping trailer, it's the first feast I've had in a long time. There's a special treat for the geese, too: grain sprinkled over a freshly picked organic dandelion salad.

Then Laura gets on her bicycle, Tom gets into his car, and I'm alone again with my little geese.

• • • • • • • • •

I SIT ON THE GRASS WITH A CAN OF BEER AND NOTICE that I still have the same peaceful feeling inside me that I had at the airfield. Is this the new me?

Gloria waddles over, checks out my bare feet, and is a little disappointed that they don't have any shoelaces. After she's subjected my toes to an extensive nibble, she settles down under my thigh and begins her sleepy-time whistle. The sound has an almost hypnotic effect on me. I shut my eyes. But only for five seconds. Then I feel a bill at my eyebrow. It's Maddin, who's usually very careful, so I don't stop him. My eyebrows and hair could do with a good cleaning anyway. I don't know what vermin Maddin picks out of my hair, but after a while he cuddles up to my head.

I'm always so happy when I'm up close with these birds. It would, of course, be odd and somehow off if I were to whisper to Maddin right now, "I love you, Maddin," so I don't. People would probably finally find me certifiable if I did that. I can just see myself in a white-walled room devoid of sharp objects sitting opposite a psychiatrist who's asking me with the utmost patience, "Mr. Quetting. You call yourself 'Goose Michael,' is that right?"

"Yes, that's right."

"And you say you love your geese? Could you explain that one more time?"

I'd rather not, but my feelings for the geese really do tend in this direction. I feel Maddin's soft feathers and his breath right by my ear. The strange thing is that not only do I sense that I love the geese, but I also sense that they reciprocate this feeling in some way.

I consider how animal conservationists often appeal to our emotions to promote their cause: "Help the animals. They're so adorable and sweet and pretty." But that's from a human perspective. The animals are here to meet our needs. But what if we could enrich our lives so much more by meeting theirs? Perhaps that way we could renew our connection with nature, which in the end would make us much happier. Is that such a crazy idea?

• • • • • • • • •

The fully automatic incubator with its hatching drawer.

Bodily contact helps develop those vital bonds.

Taking a rest in the undergrowth.

Quiet moments are special.

A family portrait.

Swimming is tiring: it's Nemo yet again who can't get enough of the water.

Keep up, kids!

The dude is messing with us.

Follow the leader.

Use your wings!

Unusual facial care.

The flight feathers are still in their blue keratin sheaths. It won't be long now.

Going through the data on the big screen at the University of Konstanz.

Nils gets his data logger so he can fly in the service of science.

The shoelace game.

A cool shower is the best remedy when it gets too hot.

Proud papa after landing.

Paul as passenger.

Paul after climbing to five thousand feet.

Finally flying in formation with my geese.

Who will be the leader?

Headed for home.

Out of the way. I want to land!

Soaring as high as the clouds will allow.

Peace of mind.

THE NEXT MORNING, WE'RE ALL EATING GRAIN, THOUGH I must say that I'm eating mine squashed—in the form of rolled oats in my muesli. The seven little geese are brimming with energy and loud. There's jabbering and flapping going on all over the place. Then suddenly, they all get very quiet and seem to be listening to what's going on around them.

Not far away from us, there's a rustling in the branches at the edge of the wood. I slowly and carefully reach for my spotting scope and spy a fawn venturing boldly out from the safety of the undergrowth onto the wide expanse of the mown field. The wind is blowing in the wrong direction, which is why it doesn't pick up my scent, which by now must be quite rank.

The fawn stands there, listening to the world with its furry ears and looking out at it with its huge dark eyes. Its brown fur looks so close through the scope that I feel as though I could reach out my hand and touch it. It's dappled with white spots.

Fawns can't keep up with their mothers all day, so they often lie for hours among leaves in the undergrowth. Their white spots and brown fur camouflage them perfectly here and they positively melt into their surroundings. How cautious and wary it looks standing there on its long legs. It has no idea that I'm watching. Unfortunately, I don't have a camera attachment on my scope yet, or I'd take a photo before the fawn disappears.

"Ssssh! Keep quiet!" I hiss at the geese as I look around carefully.

The camera is in my backpack at the other end of the bench. I get up off my backside quietly and carefully step toward it. Without taking my eyes off the fawn, I slowly reach forward. Then I suddenly stumble over Frieda, who's asleep on my other foot. The bird leaps up with an earsplitting honk and dashes under the camping trailer. I lose my balance and am frozen for a split second in an acrobatic posture—one foot under the bench, the other sticking up in the air—before I crash-land into goose doo-doo on the wooden deck, swearing loudly. With my forehead now covered in goose shit, all that's left for me to admire is the fawn's white rear end as it bounds away.

Seven Things the Geese Unfortunately Cannot Do

1. Laugh
2. Talk
3. Frown
4. Watch other animals through the spotting scope
5. Check off their sex on a form
6. Teach people how to fly
7. Defend themselves against people

• • • • • • • • •

I WIPE THE GOOSE SHIT OFF MY FACE, CALL TO THE geese—"Don't worry, I'm fine!"—and sit back down on the bench under the awning. The fawn may be gone, but there are plenty of other animals I can watch instead.

I haven't seen Jürgen the dog much lately, but there's that thieving little mouse, Fridolin, who's right now making off with another large haul of grain. I wave my fist at him, but I don't get my hopes up that I'll catch him today. The little beast is far too nimble.

The great tits in the nest box on the neighboring tree are still there, as well. Their chicks hatched a while ago and are now squatting hungrily in the nest. Ilse doesn't need to sit on her eggs anymore, so she joins Horst flying around all day catching vast quantities of minuscule pests to feed her offspring.

In this respect, at least, Ilse and Horst are quite progressive—they divide parental duties equally and fairly between them. Actually, that's not quite true. Horst very likely sits out every second or third feeding, preferring to hang out somewhere in the trees with his great tit pals and belt out songs. He might even spend time dragging thick twigs from one tree to another for no reason other than to show how strong he is.

• • • • • • • • •

IT TAKES FOUR DAYS FOR THE TEST RESULTS TO ARRIVE from the lab. There's an email in my inbox from my colleague in Heidelberg. The subject line reads: *Goose sex*. I'm a bit nervous as I open the attachment.

I blink when I read the report. What? That can't be right. I request confirmation and promptly receive photographic evidence of the results. The results state that we are a male-only group. Nothing but ganders. Not a single female. Not even affectionate, sweet, canny, cuddly Paula. Not even Gloria. But I was right about Frieda. She's stubborn Freddy, not stubborn Frieda, and has a problem with authority figures—in this case, me. The goose simply has to object to everything I propose, and that probably mostly has to do with Freddy's own insecurity.

Paula, then, is sweet, vulnerable Paul. And Gloria is not the eldest sister but the eldest brother. Should I call the goose Glorius or Glorio now? Glorius sounds a bit like a monk escaped from a monastery. I decide on Glorio.

Glorio shows no trace of the insecurity that bothers Freddy so much. Perhaps that's because he's the oldest in the group. He's the big brother to them all. Directly after hatching, the two of us had the closest contact, so we're bound by an especially strong connection. Glorio has the authority of a firstborn and, along with Nemo, the group has accepted him as their leader. He doesn't have to prove anything to himself or to others, and he doesn't have to defend his claim. No one can take away that first night he spent alone with me under my sweater. Perhaps that's why Glorio radiates such gentleness.

10

Flying at Last

WE'RE STANDING ON A GRASSY HILL LOOKING DOWN ONTO the aviaries. A northeast wind is blowing at about six miles an hour and there isn't a cloud in the sky. A buzzard circling overhead utters sharp cries, but raptors have long ceased to be a danger. The geese are now too large and, more importantly, too heavy to register on the buzzards' radar as potential snacks.

The conditions are perfect. The weather is warm, but not hot. The wind is blowing in our direction, which is just what we need to get some lift under the birds' wings. What better way to spend a day like today than by teaching your children how to fly?

A green woodpecker hammering on a nearby tree provides a dramatic drum roll to set the scene. Let's begin!

THE GEESE ARE RELAXED. JUST MOMENTS AGO, WE WERE splashing around behind our private dam. The water level in our stream is really low at the moment, and it hasn't been deep enough for the geese to dive and have a good swim. To fix this, I dragged three old wooden tables out from behind the camping trailer and rammed them into the stream bed in front of the bridge. Ta-da! Our very own dam. The water level behind the dam soon rose by four inches, and the geese understood immediately that the little stream had been transformed into a little lake.

Nemo inspected the accumulation of water from the bridge, waggled his little tail, hopped from one foot to the other, and jumped off. Descending from a height of almost three feet, he hit the water with a loud splash, landing among the other geese, which led to a full-on water fight. I would never have believed that geese could behave this way. But who ever said geese don't like to have fun?

The geese dive in like feathered torpedoes and beat their wings wildly. Sitting on the bridge, I soon got soaked by the spray. I was so wet that I decided to jump in fully clothed, and there I was, early in the morning, splashing around boisterously with the geese. I felt energized, not only by the cold water but also by the little birds' infectious, unequivocal zest for life.

I'm not saying that the geese are always cheery and in a good spirits, but when they are, there's nothing false or fake or contrived about the way they're feeling. In those moments, they are utterly and completely happy.

UP ON THE HILL, I BEND DOWN TO INSPECT THE BIRDS' flight feathers. The blue sheaths have all disintegrated, and the feathers overlap each other like wafer-thin, flexible roof tiles. The feathers are unbelievably soft and yet completely waterproof. When I stroke them, I can hardly believe that this impressive plumage has developed from the goslings' yellow fluff.

The different types of feathers on geese—the softer covert, or covering, feathers and the longer flight and tail feathers—all work seamlessly together to create a plumage that's insulating and warming, light enough not to weigh the birds down, yet strong enough to lift them into the air. It's absolutely astounding how varied bird feathers are and how perfectly adapted they are to the needs of the birds that grow them. The flight feathers of owls, for example, are particularly soft. That's why an owl—in contrast to a swan, which has much sturdier feathers—can glide through the air making almost no noise at all.

I carefully lift and partially extend the wings of each goose. As far as I can see, their feathers are all as well developed as Calimero's. That means that, theoretically, all of them should be capable of taking off.

"You can make it!" I tell them. "You can do it! You'll see!"

I think back to my first labored attempts at flight in a hang glider. When I began, flying was mostly arduous endurance training. For the first short hops, I had to keep dragging the hang glider and harness up the training slope.

On the one hand, this strengthened my muscles, and on the other, it made me painfully aware that human beings simply aren't made to fly and must go to great lengths just to get off the ground. Despite that, flying is one of the most beautiful activities we can engage in.

The geese are lined up behind me. I look at the meadow below us, stretch out my arms, and call to the geese.

"So, my dears! Come, geese, come, come, come!"

I honk the horn loudly and run down the slope, waving my arms as though I were being chased by a swarm of bees, a rowdy gaggle of seven adolescent geese trailing behind me.

I have to be careful not to trip over my own feet, especially when Calimero, Freddy, and Glorio fly past me a few feet off the ground, necks stretched forward and wings fully extended. My geese are flying. They're really flying!

• • • • • • • • •

THERE'S AN IMPORTANT FACT ABOUT FLYING THAT, AMAZING as it sounds, it's easy to forget: every flight ends in a landing. Landing is completely unavoidable. And, in point of fact, you should be more worried about landing than flying.

At flying school, when I made my first real flight at altitude with my hang glider, I recognized with great clarity the force of this rule. During training, I had mostly thought about takeoff and what it would be like up in the air. Once I was up, it was absolutely beautiful and scary at the same time. The air felt soft, smooth, and safe. The ground, in contrast, was hard, unrelenting, and deadly. I glided through the sky, saw power lines and treetops approaching, and realized that taking off and flying away were overrated. What was much more important was how you were going to get back down again.

I was completely alone up there. The flight instructor at the jumping-off point had forgotten to turn on my radio. Now, no matter how loudly he yelled from the landing spot below, he couldn't reach me. I held on tight to the control bar, not sure what I should do next.

Every landing, we had learned, can be divided into three stages: flying downwind, flying a base leg at right angles to the direction you want to land in, and then flying the final upwind leg to land, which means that to land you simply fly three sides of a rectangle, losing altitude all the time. The exercise I had reviewed numerous times in flying school now suddenly seemed fine in theory, but how was I going to put it into practice?

Theory is something that animals neither understand nor need. There is no such thing as theoretical flying for geese. No mother goose explains to her children what they should do when they fly. She doesn't draw an aerodynamic arrow in the sand with her bill, she doesn't trouble herself with illustrating her teaching or offering a variety of educational tools, and she doesn't assign homework to make sure the lessons sink in. She doesn't give any lessons in flying at all.

A mother goose trusts that her children basically already know how to fly—the coordination of movements when taking off, flying, slowing down, and landing are all innate in geese. Once their flight feathers have developed, geese can fly. They just don't know it yet, because they've never tried. You could say that they simply lack experience. Therefore, the mother goose has one critically important task. She must take her little ones by the hand (or rather, by the wing) and instill confidence in them. A mother goose doesn't explain anything to her children, and she doesn't warn them about anything either. She's simply there for them when they fly.

Sometimes I feel that this kind of hand-holding happens too rarely in human education. When we try to protect our children from all risks, we don't allow them to gain the experience they need to deal with them. And doesn't fear escalate the more you obsess about the theoretical possibilities of what might go wrong? Instead of making us feel safer, are insurance companies actually making us feel more afraid? Geese don't think about any of these things. Mind you, it's much easier for them not to worry. After all, there are no greylag geese that earn their grain as insurance agents obligated to constantly warn others about any and all possible risks.

That's not to say that the first experiences of flying are risk-free for the geese. For example, a goose could be too impatient when it flaps its wings, start its first flight too fast, be unable to slow down, and accidentally fly into the side of a house. A flying goose could also be afraid to land, stay up in the air, and lose contact with its family.

Something like that happened on my first solo flight in the hang glider. I was hanging in the air rigid with fear, gripping the control bar so tightly that you could probably have cut through the support harness and my horizontal position under the Airwave Calypso would have stayed exactly the same. In that situation, I knew exactly what landing looked like and how it had to be done. The wind conditions and visibility were within normal range. What I was missing right then was simply confidence. The belief that I could put into practice what I had learned.

Then I noticed that my flight instructor (whose motto was that the first thing to do when things heat up is light a cigarette) and the other students were forming a line on the landing field. I lost another ten feet or so in altitude before, with a cry of relief, I caught onto what they were doing. They wanted to show me from the ground how I should fly my landing pattern. When they formed their line at right angles to the runway, I was to initiate my base leg. And so they guided me down, until I could finally engage my legs and run over the wonderful, safe, beloved grass as soon as I felt it under my feet.

• • • • • • • • •

CALIMERO, FREDDY, AND GLORIO MAKE IT LOOK SO EASY as they beat their wings and overtake me. I think of Nils

Holgersson's flight and perhaps, for a couple of seconds, I really believe that if I just stretch out my arms, I too will soar aloft. Or perhaps I can fly high up into the air on Glorio's back, leaving all obstacles behind me. Anyway, I completely miss the fact that the sloping meadow has come to an end and the row of aviaries has begun.

"Stooop, geese. Land! Come, come," I yell.

Thanks to the fact that my feet are firmly on the ground, I pull up by the wall of one of the aviaries just in time. Unfortunately the same cannot be said of my flying geese. All three land together on its ample roof, many feet above the ground. Calimero and Glorio stay sitting up there, while Freddy, swept along by a gust of wind, is blown a bit farther—polishing the roof panels to a high gloss with his downy bottom, until he runs out of roof and disappears, honking loudly, somewhere in aviary row. I can't see where he's ended up, but I know that he could easily hurt himself on the gravel pathways.

"Freddy!" I call. "Where are you? Freddy!"

The remaining four geese, who never made it off the ground, don't understand what all the fuss is about. They are much more interested in the juicy grass in the grounds around the aviaries. I race down the main pathway between the aviaries and imagine finding Freddy bitterly disappointed and with his feet all broken.

I was against it all along, you know. I knew something like this would happen.

I can hardly believe my eyes when he comes waddling around a corner looking completely relaxed and gabbles at me.

Hey, old man, what are you so excited about? That was epic. Shall we do it again?

"Freddy!" I pant at him. "I'm so happy nothing happened to you. Let me take a look at you. Come here!"

I try to pick him up to comfort him after his little scare, but he just takes a dump at my feet and dashes off to the others on the grass, where he's greeted with friendly gabbling. Meanwhile, Calimero and Glorio are still perched up on the roof. How am I going to get them down? A ladder won't be much help, as the roof is too thin to hold my weight.

And so I honk the horn and call, but to no avail. A moment ago it looked as though the two were proud of their wings, but now they've no interest in using them, even briefly, to glide down from the roof. Like two stubborn crows, they remain sitting up there, looking down. Talking to them nicely doesn't work at all.

I just don't understand him. First he makes a huge fuss about us flying up in the air, and then he gets upset when we don't come down. Do you get it?

Nah. But now we'll see who's the real boss.

I resort to a laborious and somewhat ridiculous method also used to entice cats to come out from under the bed. I stand on a ladder and sweep a long broom wrapped with red-and-white tape to make it look scarier back and forth—just as Laura arrives on her bicycle.

"Micha," she calls up to me. "Have you finally gone totally mad?"

"No, I'm just doing exercises with my broom."

"I thought maybe you were teaching the goslings how to do tricks," she says, laughing.

Five minutes later, she, too, is up on a ladder with a broom, talking to the two geese.

"If you don't come down right now, you're going to have to spend the night up on the roof, and then I won't be able to look after you," I explain.

But logic helps as little as food, curses, or scaring them off. I try threats, instead.

"Fly down right now or I'll invite you over for Christmas, and it won't be as guests!"

Laura supports me with ultimatums of her own.

"I'm counting to three!"

And with enticements.

"There's super-tasty grain down here and super-delicious dandelions!"

But when a goose doesn't want to do something, he won't. Geese haven't been habituated over generations to listen to people and obey them. They're wild animals and don't have to submit to anyone.

I'm well aware that it's my fault that the two geese are sitting up on the roof in the first place, but I still find their behavior annoying. And because I can't reach the geese right now, I have to take my anger out on a smaller animal. When I spy the mouse Fridolin next to the camping trailer, once again making off with gosling starter feed, I race after him in the somewhat unrealistic hope of catching him this time.

Although the mouse is much faster and nimbler than I am, it doesn't just scamper away over the grass. Instead, it stays put and stares me down. Fridolin is clearly completely unimpressed by a big dangerous person like me. I approach the little chap slowly and bend down in front of him. He's either so perplexed or so trusting that I succeed in grabbing him with a lightning-quick move.

"So Fridolin, I've got you at last. Just stop eating the birds' food, dammit! Understand?" I whisper.

Of course, I don't seriously believe this speech will have any effect. I'm simply stressed out and now the little mouse has to suffer the consequences. But instead of apologizing, Fridolin knows exactly what he has to do: bide his time, keep still, and then, at just the right moment, bite.

When the field mouse's tiny teeth prick my index finger, I jump a couple of feet into the air and put the mouse back down on the grass—not very gently, I have to admit. Then I hop around in pain, and Fridolin just sits there and looks at me defiantly, as if to let me know he knows he won. The geese watch the whole scenario from the roof and don't even budge.

In the end, Laura and I drag over a large wooden plank and a ladder. I carefully push the plank onto the roof, checking the position of the metal struts on the aviary roof multiple times. Calimero looks at me skeptically when I test my plank bridge with one foot. Nothing cracks and nothing breaks. It's going to work. I test it out with my other foot. The corrugated sheet metal groans. Like a burglar caught in the sweeping beam of a halogen spotlight, I pause, a ridiculous grin on my face.

I put one foot in front of the other, still unsure whether the plank bridge will really hold my weight. The rest of the geese watch me with a mixture of curiosity and malicious enjoyment. Fridolin is certainly stashing away our grain as calmly as he pleases right now.

I imagine that landing on the floor of the aviary from a height of more than ten feet will be unpleasant and somewhat painful. I'd prefer to crash into a barn from the sky after a marvelous flight than onto the concrete floor here just because Calimero and Glorio are too comfortable to spread their wings even briefly.

I poke the red-and-white broom in Calimero's direction, but he simply hops a bit farther away. Then he sits down again, as though he had no wings, as though he isn't the Rambo-goose who was the first to take off, but just an overgrown hamster that has no idea how it ended up on this roof.

"Please, Calimero," I beseech him. "Please don't make things so difficult for me."

Out of the corner of my eye, I spot the other geese setting off and waddling toward their food trough. It contains the same run-of-the mill feed we've been using to bribe Glorio and Calimero for hours. When the gang passes the two contrarians, Calimero and Gloria stretch out their necks in concert, flap their way from the roof of one aviary to the next as though there's nothing to it, and finally glide down to the ground.

With my broomstick in my hand, I think: *The geese aren't just playing with me; they're making me look like a complete fool.*

11

Greedy Guts

ONE WEEK LATER, THE GEESE ARE FLYING. JUST LIKE THAT. Without any great fanfare. As though they've been doing it all their lives. When I walk slowly over the field with them following me in single file, and then I suddenly start to run while honking the horn, they take off, fly up in the air over my head, and land about ten feet in front of me.

It's fascinating to see how effortlessly they launch themselves into the air and how quickly they can come back down to earth. They flap on the spot for a moment to brake, extend their feet forward, and land elegantly on the ground. Although in theory geese fly like planes, they land vertically like helicopters. They have no need for hundreds of feet of runway.

There's only one goose that isn't flying: Nemo. I think he's just too fat. He really tries. He races along madly behind the others, flapping his wings and gabbling loudly. But while the others soar up into the air and quickly gain altitude, Nemo gives up with a heartrending honk and stays on the ground with me, desolate.

"Nemo, don't be sad," I say as I try to console him. "You'll get the hang of it soon! I'm going to give you one-on-one training. And I'm putting you on a diet!"

I'm hoping that Nemo's not going to end up in that vicious cycle we can all relate to in some form from personal experience: because he's overweight, he moves less, and because he moves less, he stays overweight. Although, in this case, I don't know if we can really talk about being "overweight." Don't geese have an inner sensor that prevents them from eating too much? I've certainly never heard of a body mass index for geese. I also don't know what a real mother goose would do in this situation. Would she withhold grain from her babies?

I wonder if there are overweight geese in the wild that rationalize their failure to get off the ground by explaining that flying has never interested them and the best dandelion fields are within walking distance anyway? Nemo doesn't seem very keen on my proposal.

A diet? You've got to be kidding! I'm only three months old! I'm still growing, for heaven's sake!

He gabbles at me, turns around, and waddles sulkily to the nearest tuft of dandelions.

● ● ● ● ● ● ● ● ●

A THIN VEIL OF MIST IS HANGING OVER THE RUNWAY, struggling to survive against a strengthening wind, when

Laura and I arrive at the airfield the next morning with the geese. Even though chubby Nemo can't take off yet, we want to try flying with the ultralight for the first time.

We can't cater to Nemo any longer, because it's been a long time since the weather conditions have been as perfect as they are today. And because this is a research project, we need to make at least some effort to keep to the schedule. I'm also hoping that the other geese might spur Nemo on to dig deep and find the extra impetus he needs to take off.

Today the wind is coming from a northerly direction, so it's aligned nearly perfectly for runway 01. In pilot speak, the direction of takeoff is given in the degrees of the compass, and for simplicity's sake, pilots leave off the last zero. That means that a takeoff to the south is described as 18, a takeoff to the east as 09, and a takeoff to the west as 27. In our case, 01 means we'll be taking off in a northerly direction.

The geese are now so familiar with the Atos that they begin to gabble loudly as soon as they see it from afar. They sound as though they're greeting a somewhat overgrown, misshapen member of their goose family. I'm pleased the conditioning has worked so well. Instead of seeing the ultralight as a monster, the geese associate it with pleasure.

I ready the plane for takeoff and go through the all-important checklist. It sounds boring to check items off a list, but it's one of the major tasks when you fly—as it should be. There's no other way to disable the potentially fatal influence of habit. The level of tension that duly dogs you on your first solo flight makes you automatically check everything five times. This tension is a completely appropriate physical reaction, because it keeps you from taking unnecessary risks. A healthy dose of awe is absolutely desirable.

Unfortunately, after a while, even something as awesome as flying begins to feel familiar, which pushes the necessary sense of awe increasingly into the background, until at some point you hardly feel it. You catch yourself getting careless with vital checks and ignoring some of the rules. In this situation, working through checklists thoroughly and rigorously is the only thing that ensures your safety, even when you've operated the controls a thousand times before. When procedures become routine, flying can become deadly. I've lost hang-gliding colleagues simply because they forgot to harness themselves to their machines before they took off.

I check the charge on the battery, test the different flap positions, and turn on the flight instruments. External conditions couldn't be better—but I still have an irrational fear that something bad could happen on the very first flight.

We unload the crate of gabbling geese from the van and carry it over the grass. Then I climb up into the cockpit, press the ignition knob, open the throttle, and taxi slowly toward Laura and the geese at the start of the runway. The geese in the crate are now directly under the right wing of the plane.

I take one more deep breath, because now things are getting serious. The breeze coming through my open helmet blows gently over my nose, and my body tenses slightly as I press down on the ignition knob again and hear the penetrating warning beep of the engine controls. As a gust of wind sweeps across the runway, I have to use my shoulders to stabilize the forty-foot wingspan, and the right wing comes dangerously close to the crate full of geese. The green ignition light blinks. Everything is ready. Nervousness spreads through my system, and even Laura's expression shows signs of strain. The geese, however, are happily gabbling away to themselves.

• • • • • • • • •

I LOOK UP ONE LAST TIME AND CHECK THE DIRECTION OF the red cotton telltales that let me know which way the wind is blowing. They're blowing directly toward me, which means the wind is coming from dead straight ahead.

"Now! Off we go!" I call, and Laura opens the doors of the crate.

I've attached the horn to the right-hand side of the control bar, and now I squeeze the black bulb with all my might. The honking of the horn blares out over the airfield, the geese storm out of the crate, and I open the throttle as far as it will go. The motor roars. A jolt pushes me back into the pilot's seat as the plane lurches forward.

We race down the runway together. Nils is the first to fly, and then Glorio launches himself into the air, and finally all my geese are gliding along next to me with exhilarating ease. The plane, however, is not yet going fast enough to take off, and it's still bumping along the runway. I find myself surrounded by beating wings and rhythmically moving necks. The geese, my goslings—who were tucked inside their eggs just a few short weeks ago—are now flapping their wings alongside me.

Suddenly, all becomes quiet as the plane loses contact with the ground and softly climbs aloft. We're flying together for the first time.

"Yippee!"

The sound escapes from deep within my chest, and the geese answer with a loud honking that rings out over the airfield. We've freed ourselves from the shackles of the earth, gravity doesn't mean anything to us anymore, the sky is ours to explore.

I don't have much time to indulge in euphoria, however, because we're not flying very high, and I have to quickly focus on what I'm doing. I mustn't catch up with the geese in front of me, and so I fly as slowly as I possibly can about six feet above the grass runway.

The runway at our small airfield is only about 250 feet long. Therefore, our short excursion is over after only about a minute, and I have to think about landing. I carefully reduce speed, the geese avoid the wings as they, too, slow down and drop back, and a few yards farther on, I land softly on the grass. I coast to a stop with the geese flying behind me, and finally they land right beside the plane. We've passed our first little flight test.

No sooner am I back on the ground than I rip my helmet off my head and crawl to the geese on all fours with my head bobbing with excitement.

"Weeweeweewee! You're just great!" I call. "I'm so proud of you!"

I can barely get the words out, I'm so happy. But then I notice that only six geese are present for our round of mutual congratulations. One goose is missing. In a state of mild panic, I check the colored leg bands and know immediately who it is. Nemo. He didn't fly with us. He didn't make it up into the air. Did he veer off in time? Or have I run him over?

I look down the runway, but there's no goose to be seen anywhere. The geese and I race over the airfield toward Laura, who's still standing at the beginning of the runway and is now waving her arms. She's almost half a mile away. I've no idea what she's trying to communicate by waving her arms, and I forget that I could just call her. But then my cell phone rings.

"Have you seen Nemo? We lost him on the runway," I pant.

"No, but I noticed through my binoculars that one goose didn't take off with you. There were only six geese in the air. I'll drive around in the van right away and see if I can find Nemo and scoop him up."

The geese and I watch from a distance as Laura climbs into the van and drives toward us along the runway. There's a smile on her face when she gets out.

"I've found him!" she calls with relief in her voice.

"Where is he, then?"

"Guess!"

"No idea."

"Well, it's a no-brainer. He's in the field full of dandelions right over there. The little greedy guts is calmly stuffing his face while we're rushing around like the fire brigade looking for him."

• • • • • • • • •

IN THE EVENINGS, I SIT ON MY BENCH UNDER THE AWNing in front of the camping trailer. It's the style of bench you often see set out by folding tables in beer gardens and outdoor cafés. I relax while the geese enjoy eating grass. After having sat for such a long time on my beer-bench observing the geese, I'm pretty certain that I've trained my bench-sitting muscles so well that I could outsit each and every one of the patrons at Germany's famous Oktoberfest.

While I eat another of Heinrich's sandwiches for dinner and watch the geese graze, I eye Nemo suspiciously. Instead of lying under the bench in shame, he's happily shoveling in grass with his bill. It doesn't look as though he's going to hold back or watch what he eats just so he can get into the air next time.

And why should he? Why should a goose be interested in dieting? Nemo simply eats whatever he wants to eat because he knows: one day he's going to be able to fly. Because the geese—unlike geese in the wild—get an unlimited supply of high-calorie grain, I still have doubts about whether he'll ever be light enough if he keeps on eating like this, but I also believe that it's only a question of time before this situation works itself out and he, too, takes off. It occurs to me that Nemo doesn't have a problem. It's me with my expectations who's dreaming up the problem in the first place.

Seven Things Geese Absolutely Cannot Do

1. Lie
2. Network
3. Make tactical power plays
4. Pretend
5. Apply psychological pressure
6. Grandstand
7. Climb the career ladder

● ● ● ● ● ● ● ● ●

AS THE EVENING SUN SHINES FAINTLY OVER THE FIELD, I notice another greedy guts who's stealing from us once again. Fridolin is scurrying along his little tamped-down path, his cheeks full of contraband. A few days ago, I followed his tracks. They lead to a hollow tree stump, where the mouse has piled up a whole mountain of grain.

For Fridolin, the mountain of grain is probably an all-you-can-eat paradise. It's way more than he can eat in a winter, but it's not that much when you think about how much grain seven geese go through. And so I decide to leave Fridolin alone for the time being. I've no desire to get bitten again or to get angry with the mouse. You're welcome, Fridolin. I wish you all the best. He might as well enjoy the grain. After all, there's a limit to how much a little mouse can haul away.

I'm grinning somewhat apologetically as Fridolin trots away, when something unbelievable happens. No more than fifteen feet away from me, out of the blue, a kestrel plunges down from the sky, grabs the mouse with both feet, and disappears in a flash back into the sky with its prize. A flap of wings, a squeak, a shriek, then silence. It's all over for Fridolin. Paul is the only one of the geese to glance up at the kestrel. I've been sitting under the awning for a considerable length of time, so the kestrel couldn't have spotted me from the sky—if he had, he would never have taken the risk.

Suddenly, my little skirmish with the mouse seems petty and overblown. Why hadn't we gotten along better? Why hadn't I behaved more magnanimously toward Fridolin? Why had I never invited the mouse over for a couple of kernels? It's too late for all that now. The food snatcher has himself been snatched from the field as food. Well, I think, I'm just happy that so far all the geese have survived. There's no such thing as overeating in the wild. Just eating or being eaten.

That might sound cruel and heartless, but I see it differently. It is the harsh realities of nature that have kept me grounded over the past few weeks. The geese are here, they need me, and I have no choice but to get involved. That's just the way it is. If I accept this, I'll be fine. Before I was with the geese, I rebelled a lot more. I fought things: my separation from my own children, all the complications of life that appeared suddenly and then didn't disappear just as suddenly.

Despite the fact the geese have no interest in teaching me anything and couldn't care less about my private life, I have learned something really important from them: radical acceptance. How to abandon myself confidently into the clutches of life. Things are what they are, and that's as it should be. It's only when you focus on the basics that you can really live in the here and now. Sure, you could say, you might just as well have talked to your grandma to learn that lesson, but hearing words of wisdom is not the same as actually living them.

Nemo is perhaps a bit on the heavy side, but I'm finally certain that everything will work out. Little mouse Fridolin: adieu.

12

Freddy Veers Off Course

I HONK THE HORN LOUDLY TWICE AND YELL, "NOW, Laura!"

Laura opens the crate off the end of the wing, I push the accelerator lever as far forward as it will go, and the geese rush out gabbling loudly. A VW Touareg is racing along parallel to us on the airfield like an enormous black panther. I'm pressed back into my seat, and the brisk head wind lifts us up into the air after only a dozen or so feet.

Our small airfield has been transformed into a film set. At first, it was the local papers taking an interest in what the geese were getting up to, and now the geese are appearing on TV. There's a large high-res camera attached to a vertical contraption on top of the Touareg, and there's a camera operator inside the van who can point the camera in any direction using two joysticks.

A crew employed by the German public TV station ZDF is making a film about migratory birds, and they want to film me taking off with the geese. The whole thing is an enormous undertaking and seems highly professional, but just now I have had to sit in the sauna that is my cockpit for half an hour wearing all my gear, waiting with the whole crew until someone realized that somewhere a plug wasn't plugged in all the way.

The geese extend their wings and fly along close to the plane. For the very first time, we're flying in a V formation. This is the moment I've been dreaming of for so long. I'm part of a flying goose family. I steer the plane and look around enthusiastically. The geese are moving their wings up and down next to me, hanging so effortlessly in the air that it looks as though they're floating.

The V formation is the most energy efficient for the geese, because in this formation they can use the turbulence coming off the bird in front to help power their own flight. You can make this air turbulence visible on planes using colored smoke. You could do this with geese, as well, but of course that isn't necessary. You don't need to explain this energy-saving trick to the geese. They just know that the V formation is the best one for them to use. They don't need to try out a T, B, or X formation first.

The wonderful thing about the V formation is that each goose is both dependent on the goose in front of it and needed by the goose behind it. We're all pulling together; we're all flying in the same current of disturbed air; we're all pursuing a common goal.

Perhaps that's why the sight of a formation of wild geese in the sky moves us even when we watch them from afar. We're standing below among people and we can sense that the geese up there are forming a team in which each member supports all the others, and at the same time we know how rare it is for people to do this. We really should spend more time traveling with family. It feels so good.

I count off the flying geese and come up with seven. Wait a moment. Did I say seven? That means that Nemo must be up in the air, as well. I check off the colored bands on the birds' legs and there, indeed, is Nemo's blue one. Nemo is flying with powerful, elegant wing beats right up front. He did it. My chubby little boy is flying as the lead goose at the tip of the formation.

Nemo looks over to me with his dark eyes. I can just imagine what he's thinking.

Did someone just say chubby? Did you call me a greedy guts? From now on, I'm the boss here! You'd better bundle up if you're going to fly up high with me! I told you I was just growing.

I'm so thrilled with Nemo that I completely forget something important. I'm running out of runway to land on. We're only filming the takeoff, so the plan was to take off briefly and then land again without having to make a turn, but now we've gone too far, and I need to turn to avoid crashing into the trees.

Flying low, I bank gently to the right. With a wind this strong a 180-degree turn is not enough, because we can't land safely in the opposite direction to the one we took off in. And so we'll have to make a small circuit around the airfield to get back to the approach I need for landing. The geese aren't yet used to flying for so long. They're not sweating, of course, but I think I can tell that it's an effort for them. Throughout the 360-degree turn, I keep honking the horn to keep the geese close to the wing.

"You can do it!" I call. "Come, Nemo, come! Fantastic."

Suddenly, I hear a loud honking that I know only too well. It's Freddy, and when he honks, things never turn out well.

Freddy's flying toward the back of the formation; therefore, I can't see him right away. It's only when he breaks away from the group and dives down in a tight ninety-degree turn that I realize what he's doing. It appears that, once again, Freddy doesn't want anything to do with us and insists on doing his own thing. All alone, he quickly loses altitude. I keep visual contact with him for a moment and try to make a mental note of roughly the direction he flew off in, but then I have to focus on the remaining six student pilots and the difficult approach for landing.

What's going on with Freddy—again? Could it really have something to do with the camera that's following us? Does Freddy have concerns about privacy and data protection? *He'll just have to look after himself*, I think, as we begin the approach for landing. I lower the flaps on the rear edge of the wings into position for landing and ease off the accelerator. The lowered flaps increase the uplift on the wings, which in turn allows me to fly more slowly, while at the same time losing altitude without falling too fast. As I'm attempting to stabilize the plane so that it's pointed directly at the landing strip, the remaining geese also break formation and start executing aerobatic maneuvers all around me.

Even though they're only about ten feet above the ground, they turn themselves quickly over onto their backs, fly upside down for a few seconds so that they're looking up at the sky, and then right themselves again. I'd read that geese could master this motion, but I never thought I'd observe this flight maneuver at close range. Like lowering the flaps on my wings, turning upside down allows them to lose altitude rapidly in a controlled manner. You could say it reverses the profile of the wings so that uplift becomes downthrust. In the brief time that they're flying upside down, the geese are not only falling, they're actively being sucked down.

I've not the slightest idea how the geese have mastered this maneuver. They definitely haven't learned this trick from me. I'd need a completely different plane to teach them that move, as it's completely impossible with the Atos. What I can say is that it's part of their genetic programming. But that doesn't make it any less amazing.

Around three hundred feet from touchdown, they're still flying in front of me, and I'm worried I might run them over when I land. I pull the control bar back as far as it will go while accelerating so that I can hop over the geese and get them out of the danger zone. The plane is sweeping down the runway at almost fifty-five miles an hour. Then I reduce speed and finally put the Atos down, more or less gently, on the grass.

A moment after the plane stops moving, all six geese land elegantly around me on the grass. I'm ecstatic that they've arrived safe and sound. Weak at the knees, I fall onto the soft ground. I'm surrounded by gabbling geese and smell the freshly mown grass. That's when I realize that the film crew, of course, has been waiting for a heart-warming moment just like this one, and they are now busy filming me and the geese.

While I lie there with the geese, I'm torn. On the one hand, the pictures will probably be amazing and very beautiful, and I'm happy that our project will be brought to a wider audience. On the other hand, I don't know if this scene—Goose Michael on the brink of tears because all his geese have flown with him for the first time—will seem kitschy and perhaps even a bit ridiculous when it's broadcasted on TV.

That's the problem with reports about animals. Everything always has to be moving and heartwarming, and ideally the animals are anthropomorphized like they are in fairy tales. Little wonder that films about animals are often a bit too saccharine. On its own that wouldn't be too bad if it didn't make us forget the reality of the relationship between people and animals. I don't know if it's possible to convey in pictures what it means and what it feels like to live with a family of geese for weeks. It might come across as ridiculous that I'm lying here in the grass with the geese, but it's not for me. For me, it's the logical result of the relationship I've forged with them.

I suddenly remember Freddy, and I immediately forget that the crew is still filming.

"Laura," I shriek. "We've lost Freddy! Did you see him after he veered off?"

But Laura shakes her head.

"All I noticed was that one goose left the group. It looked as though he flew toward the wood."

"Freddy again."

During recent training flights, we'd been equipping the geese with radio telemetry transmitters so that we could find them in case of emergency. But because today we'd planned just a brief takeoff for the film crew, we didn't do this. So now there's no way of finding Freddy quickly. I leap up and the geese flutter off the ground.

"Gather the geese. I'll see to the Atos," I call to Laura. "We have to set out as quickly as we can to look for Freddy!"

We climb into the van and, armed with the horn and two sets of binoculars, we set off to rescue Freddy. I have to say that we don't have a plan. How are we going to find a greylag goose that knows how to fly, in an enormous area with a wood, lots of fields, and a village?

With the windows rolled down, we drive the perimeter roads honking and calling and scanning the undergrowth with our binoculars. It takes us two hours to accept that our search is hopeless. I curse the "stubborn," "untrainable," and "rebellious" Freddy, but then I wonder whether Freddy actually wants to be rescued. Didn't he decide of his own free will to leave the group? His maneuver certainly didn't look like a mistake. Do I even have the right to "save" him, that is, to bring him back to a place where—as he's made clear time and time again—he would rather not be?

It's been dark for a while by the time we finally break off the search. But it occurs to us nonetheless to inform the owner of the restaurant at the airfield about the loss of a goose. So that if and when hunger drives Freddy back to the airfield near the restaurant, where from time to time breadcrumbs drop onto the ground between the tables, someone will call us. I feel worn out when we load the geese back into the van and drive them to their aviary. Everything looks normal and the geese look relatively peaceful, but one goose is missing. Freddy's not here.

Laura peddles home, but I sit down in front of the camping trailer and blame myself. *What on earth is wrong with Freddy?* After all, I treated him exactly the same way I treated the other geese. But is that really true? In a family of seven, there's probably always one problem child. But out in the wild, a goose that young would never have left its family. So why wasn't I more careful? Can I really blame Nemo for the fact that I was so distracted? Or is Freddy already grown up after all, which means that what happens to him is no longer my fault?

Freddy, you silly goose, why are you doing this to yourself? I wonder. Then my phone rings in my pocket. It's Carola, our ever-vigilant colleague, calling from the front desk at the Institute.

"Micha, you're not by any chance missing a goose, are you?"

"Yes, I am. We've been searching all evening. Have you seen Freddy?"

"No, but someone from Wahlwies just called. There's a greylag goose sitting in someone's front yard and it won't budge. It won't let anyone catch it. Could that be Freddy?"

"It definitely sounds like Freddy. I'll be there in ten minutes."

• • • • • • • • •

I'M NOT SURE IF I'LL BE ABLE TO CATCH FREDDY ON MY own in unfamiliar territory, but luckily I reach Laura right away. When I pull up in the van, she's already standing outside her house, eating a carrot. She jumps right in.

"How did you know it was Freddy?" she asks excitedly.

"Well, there was talk of a greylag goose with a red leg band, sitting obstinately in a front yard, calmly allowing itself to be looked at by a dozen children. That kind of narrows it down, doesn't it? It has to be Freddy."

I ignore the speed limit, and thanks to Mrs. Google, who lives in my cell phone, we arrive at the address fairly quickly. When we turn down the street, the front yard in question is immediately obvious. An animated crowd of children and adults has gathered in front of an impressive single-family home.

At first, we're taken for more spectators, and we're directed to a parking spot in an orderly fashion. I make my way through the crowd and hear someone say, "It's lost its family," before I finally set eyes on the goose. There's no doubt that it's Freddy there on the grass. He's shifting his weight from one foot to the other, looking down at the ground. He seems a bit disoriented to me, even a bit listless. He looks as though he's already accepted that he's destined never to find his family again. I squat down on the grass and approach him slowly and carefully.

"Freddy, what are you up to? Papa's here. Come on, let's go home," I whisper.

When he hears my voice, he lifts his head, looks at me, and gabbles softly. It's a sound I've never heard from a goose before. A sad, relieved but at the same time subdued peep-gabble.

Freddy doesn't flinch or flap but simply seems happy that I've come to collect him. I pick him up, which he allows me to do right away, walk him to the van, and thank the people for their care and concern. They knew of our project thanks to a newspaper article and therefore guessed that the goose that was behaving so oddly probably belonged to us.

"Did the goose really recognize you right away?" the people ask.

"Wouldn't the goose have found its way home all by itself?"

I explain the project as briefly as I can, because I don't want to keep Freddy waiting in the van for too long. I can imagine that he might have found his way home to us using just his sense of smell and direction. Perhaps he intended to be gone just a short time. But he probably hadn't considered that he'd be completely alone.

Freddy hardly makes any noise at all on the drive back. It's as though he knows deep down inside that instead of being cool and independent, he's insecure and afraid that he doesn't belong to the rest of the group. Perhaps I'll be a goose whisperer yet.

Back home, I immediately put Freddy in the grass among his siblings and witness a very loving and heartfelt reunion. The other geese waddle up to him, caress him, and cuddle up to him as though he's been gone for an eternity or barely escaped the stew pot. They welcome him openly and without judgment, even though at noon he left the family of his own free will. Even Freddy seems to thoroughly enjoy the attention he's getting from the other geese.

Yes, yes, family. I'm happy to see you again, too. I was quite concerned about leaving you alone with that dude. But enough with all the cuddling!

I watch the geese in the grass and I get to thinking. I'm as happy as the other geese that Freddy's back, but his behavior is still a puzzle to me. I'm certain I showed him just as much affection as the other geese. And yet we didn't succeed in establishing a really stable father goose–baby goose relationship. Did I somehow unconsciously communicate that Paul rather than Freddy was my favorite? Should I have hugged Freddy more, even though he made it clear fairly early on that he didn't want that?

Or was it a question of drawing attention to himself, especially during the filming? Is this exactly what he wanted: that we would worry about him, look for him for hours, and then be ecstatic to see him again? Is he trying to tell me something?

I mull things over but don't come up with any obvious answers. And that, too, is something I hadn't seen clearly before: geese are individuals, and I don't always understand why they do what they do.

13

The Perils of Puberty

IT'S SOMETIMES SAID THAT WITH CHILDREN THE FIRST TEN years are the best. These are the years before they get minds of their own. These are the years when they offer you unconditional love and, by and large, do what you ask. I don't believe this is really what happens with human children, but when it comes to geese, I'm not so sure. A few times recently, the geese have really been getting on my nerves.

After three and a half months, the goose kids are now well into puberty, and daily life with them has become a constant struggle. We're at a standstill with flying. Instead of affording me peace, companionship, and acceptance, the geese are raising my blood pressure.

Of course, if I simply followed their moods and spent the whole day chilling with them at the lake, hanging out in the meadow, and admiring their short modest flights, then there would be no conflicts. But that's not what I want. What I want is to complete our joint project; therefore, I need at least a basic level of cooperation from them. Yet at nearly every turn, they dig in their heels and noisily refuse to comply with my most reasonable requests.

Ilse and Horst, the two hardworking great tits, often sit next to one another on a nearby branch and sing songs that are probably all about how successful they've been with their brood. Sometimes I get the feeling they're following the conflicts between me and the geese with smug satisfaction.

He simply didn't set any boundaries for his chicks. Do you know he even allowed them to crawl up under his sweater?

That always happens with precocial birds—there are never any strict rules.

You have such a fine appreciation for these things, Ilse.

• • • • • • • • •

AND I HAD BEEN SO PLEASED AND HAPPY WHEN THE GEESE flew with me for the first time—that first time when we were in the air together. They're still happy every time they see the plane, and they fly along with me, but they're barely far enough off the ground to make it over the trees at the end of the runway. That's simply not high enough for the measurements we want to make. We're so close. We just need a little bit more. But when I try for those last ten feet or so, the geese put on the brakes.

Until now, we've managed flights that have lasted ten minutes at the most. That means we're basically just flying wider circuits frustratingly close to the airfield. I can't seem to get the birds up to a reasonable height. When I climb higher, the geese don't follow, which forces me to come back down again. And why should they make the effort to fly higher just so that the Max Planck Institute can take some measurements?

Their reluctance to gain altitude means that I often end up flying the ultralight much too low over the trees. Then I swear and prefer not to think what might happen if the motor were to give up right then. It's not at all relaxing to fly so low. I'd have to react to the slightest problem immediately, and I'm in constant fear of an imminent collision with the top of a tree. The geese, however, find this height just perfect. They couldn't care less about my stress or anxiety.

Since they've learned to fly, I've been gradually losing control over them. They can now fly off any time they please, and I can do nothing to stop them. I'm having a hard time getting used to that. Every time they decide to take off on their own could turn into an emergency.

IN EVERY EXPERIMENT IN IMPRINTING BY COLLEAGUES around the world until now, gradual independence from the parent bird has been a perfectly natural and desirable process. The geese were imprinted, grew up with their parents nearby at all times, and gradually became more and more independent. They flew out to a lake in the evening to sleep, and in the morning they returned to their human caregivers to be fed. Sometimes the geese stayed away for days and then suddenly reappeared. None of that was a problem.

Unfortunately, things are somewhat different in my case, because my goal is to get the geese to fly to gather data. Until we've gathered all the necessary data, I can't lose a single goose. Every single flight and every single goose provide important information that has never been collected in this way before. In addition to the reluctant geese, we're experiencing technical problems. There are delivery issues with the high-res data loggers that have been specially designed for our project, and the longer I wait, the more reluctant the kids are to participate.

The fact is the geese are no longer sweet little goslings that slip up under my sweater, waddle along behind me, and do what they're told at the slightest honk of my horn. Waddling, in particular, is a thing of the past. I still walk out in front of them, but their lordships now fly, of course. When I want to go down to the stream with them, as we did in the old days, they don't want to walk. Instead, they suddenly overtake me, flying low and swooshing past, only to wait for me at the stream with bored expressions on their faces.

It's as if they want to say: *Papa's so dull and so normal. He can't even fly.*

That's true. I can't. But I still wish they'd show the old man a bit more respect.

We don't want to take part in your stupid program anymore!

What interest do we have in your daft measurements?

We don't give a feather about the weather.

Who is this Max Plankton, anyway?

I'm used to such behavior from Freddy, but I have to say it's really harsh that recently even Paul has been rebelling against me.

• • • • • • • • •

BUT BACK TO THE BEGINNING. WE'RE SITTING, ONCE AGAIN, at that "boring" stream. The place the geese enjoyed so much in those earlier days when they could think of nothing more delightful than peacefully splashing around between my feet and gazing up adoringly at their papa or soaking him with water. Suddenly, Freddy launches himself into the air for no apparent reason. With a loud honk that probably means something like, *Bye, you schmucks!* he flies out of sight between the bushes in the direction of Castle Meadow. Maddin, his faithful companion, naturally can't bear to be without him and hurries after his role model.

The words "Freddy, you dirtbag!" slip from my lips. That might not be a proper way to address a goose, but it's not always easy to control yourself when you're around geese in the throes of puberty. I jump up and look around, but I can't see Freddy anywhere. I immediately begin to panic. I hurriedly drive the remaining five geese along the path back to the aviary, stretching out my arms and making awkward flapping motions to encourage these geese—who had been so eager to fly on the way out—to now fly on the way back so that we'll get there more quickly.

But suddenly they're no longer interested in flying. It seems they want to stop at every corner to eat dandelions, and I have the feeling they're dawdling on purpose just to give the two runaways more time. In the end, I break off two long branches from nearby bushes and drive the geese by waving the sticks about behind them. Toni and Moni, the two ruddy shelducks, comment on the goose roundup with scornful, mocking squawks.

When I get to the camping trailer, I immediately shut the geese in the aviary. Then I grab a pair of binoculars, run to Castle Meadow, and honk the horn as loudly as I can.

"Freddy, Maddin, come, come, come! Jeez, Freddy," I call.

I hope that no one's watching as I alternate between swearing and sweet-talking the geese from the undergrowth.

I beat my way through thorny shrubs, sink in moss, slip down a muddy slope, and arrive at the Institute festooned with grass, looking like I've been living in the woods for years, to report that two of the geese have flown away. Perhaps, as with Freddy's previous escape attempt, helpful people might have called in with information. I'm shaking the dirt from my clothes when I spy Maddin, peacefully slurping water from a saucer under a plant pot in the Institute's front garden.

"Maddin!" I shout at the goose. "Come here right away! What on earth do you think I've been doing all this time? And where did you leave Freddy?"

Maddin doesn't even flinch. He just looks at me a little surprised.

It wasn't me! Chill, dude!

I march purposefully back to the others carrying Maddin, drop the goose off in the aviary, and hastily return. Freddy can't be far away. I search the fields and along the edge of the wood, but I can't find any sign of him. Even with binoculars, it's difficult to spot a single greylag goose in the bushes—their feathers blend in with their surroundings and camouflage them well.

I trudge through the undergrowth, tear my arm on a blackberry bush, and almost fall into a patch of stinging nettles. I decide to cool my feet off in the castle pond. And there he is. He's swimming along the bank, hard at work digging up the roots of the freshly emerging reeds. Unfortunately, I can't reach him from land, because the ground is so muddy that I would probably sink in up to my hips.

I stand on the bank for a while talking to Freddy, but he doesn't move closer to me—not even an inch. By the time I get back from a quick trip to the camping trailer to get an old surfboard I plan to lie on so that I can paddle over to him, he's disappeared. It takes me another circuit of the pond and much peering into the undergrowth on the bank to finally spot him. I pick him up and carry him back to the camping trailer.

• • • • • • • • •

THE NEXT ONE TO HEAD OFF IS NEMO. ONE DAY AFTER Freddy's attempted escape, we're out and about in Castle Meadow when we're caught in a thunderstorm that I—even though as a pilot I can supposedly read the clouds, and always monitor weather and wind conditions—absolutely did not see coming. In just a few minutes, the sky darkens ominously and the trees begin to bend over in the wind. Before I've had time to roll up the camping mat, there's a rumble of thunder, as well. Raindrops pelt down over the meadow like intense volleys of goose droppings.

The geese waddle in the rain in front of me. Even though I'm honking the horn madly and shouting over the wind and rain to keep the geese close to me, Nemo shifts gear, speeds up his waddle, and after just a few steps, he's up in the air. In contrast to Freddy, Nemo has the confidence of the whole group, and they take off after him right away and disappear behind the nearest treetop. Only Calimero remains obediently by my side awaiting further orders, but I have no idea what instructions to give him. And so I stand by my most steadfast goose soldier with no idea what to do, and in no time, I'm completely drenched. I grab Calimero and race down to the aviaries.

Glorio, Nils, Freddy, and Maddin are already there waiting for me, at least—though Glorio and Nils are on the roof. While buckets of rain continue to pour down, I put a ladder up to the roof, fix myself arm extensions from an old pipe, and wave them around at the geese, while calling, "Come, come, come." Unfortunately, the annoyance in my voice doesn't register with either of them. They seem to take my movements with the pipe to be some kind of a skipping game, and they happily flap from one side of the pipe to the other. If the sun were shining, I'd enjoy this, but the rain is pelting down on my face, my shirt is sticking to my chest, my shoes are making squelching noises because they're so wet, and Nemo and Paul are still missing.

I jump down from the ladder, race into the storeroom, and tie a piece of red-and-white warning tape to the end of the pipe in the hope that will make the rascals show the broom a little more respect.

"Glorio, come, come, come," I call.

I brandish the pipe in the air. Then I slip off the ladder, scraping my shin on the rungs. For the first time since the geese came into this world, I'm really angry.

"Come down right now! Move it!" I roar, jerking the length of pipe back.

I don't know if it's because of the way I shouted at them or because of the red-and-white plastic flapping at the end of the pipe, but both let out an indignant honk and finally let me catch them.

But how am I going to find the rest of the gang again now? I get into the van in my soaking wet clothes and bump along the tracks in the fields around the Institute, honking my horn. But in this weather, it's futile to try to spot a greylag goose on the property. Because I want to do something rather than nothing, I drive up and down the edge of the wood, calling for Nemo and Paul, even though I highly doubt they can hear me.

The thunderstorm moves on as quickly as it arrived. All at once, the sky is cloudless once again, and it looks as though the sun has just risen. Castle Meadow is slick and shiny when I get out and sit down in it. I can't get any wetter than I am already.

"Come, geese. Call in, please," I mumble rather pathetically.

I make an effort to at least enjoy the sunshine. Do I have a right to catch the geese if they prefer to be free? But if I lose the geese, one by one, before we've even had a chance to take the first proper measurements, then all our effort and expense will have been wasted.

Then I think I hear a gabbling somewhere between the trees. What was that? I pull myself together, take a few steps through the squeaky wet grass, and stop. Perhaps it wasn't gabbling after all? I reach for my horn and honk it. I'm answered by a loud honking, really close by. I race to the trees, and after just a few steps I spot Nemo. He's sitting next to a tree, sees me from afar, and waddles toward me and my horn. No sooner do I pick him up than he's nibbling on my hair and checking out my face with his bill.

Now the only one missing is Paul. I walk back to the aviary with Nemo and make myself a coffee before setting out again. I need to be smart and rational. I have to think. This coffee tastes terrible. *Why Paul of all the geese? Why has Paul decided to join the miscreants?*

To distract myself, I sit on my bench and start fashioning new pads for the harnesses on the data collectors. It's simple manual labor, guaranteed to relax me right away. I cut the pad from soft flexible fabric using a carpet knife so that I can cushion the underside of the harness with a U-shaped piece of foam rubber. The pad ensures that the spine of the goose has room to move, while allowing the underside of the harness to lie flat. Just as I'm spreading superglue onto the foam rubber, I hear the sound of many wings beating in the distance.

A flock of greylag geese appears in the sky. I drop the superglue and reach for my horn. Then I race out to the path in the field and honk until my hand hurts. The flock consists of about eight geese making their way majestically across the sky. I honk and honk.

Then a single goose breaks out of formation, quickly loses altitude, and prepares to land in the middle of the cornfield. It's Paul. The wet clayey soil of the field sticks to his feet, and he hops rather than waddles toward me.

I stretch my arms wide. When I finally have Paul again and can hug him, I bury my nose in his neck and breathe in the scent of his feathers. I simply can't get enough of this smell. Right at this spot, the geese still smell like little goslings. A while ago, the whole flock smelled like this, but that time is now gone forever.

• • • • • • • • •

LUCKILY, GOOSE PUBERTY IS NOT ALWAYS NERVE WRACKing. When we're all lying together again in the shade in Castle Meadow, I try a small experiment. After we've relaxed, the birds pull impatiently at my pant legs to let me know that it's time for a nice bill-full of grain back at the aviary. Up until now, I've always jumped up right away and led the team home. This time, however, I want to see what happens if I just stay calmly sitting there.

For a few minutes, the geese appear to be frustrated, but then they organize themselves. Nemo steps up to the plate and honks to his siblings.

The old man might want to keep lounging around, but my stomach's so empty it's almost dragging on the ground. Follow me, everyone!

Nemo starts moving off in the direction of the aviary. What's interesting is that he doesn't take the direct route over the open meadow but leads his geese under the cover of the trees at the edge of the wood, all the while keeping an eye on the sky and the surroundings. The other geese trust Nemo and follow him. It makes me really sad that

Nemo can so easily take my place, but at the same time, his decisive action fills me with satisfaction. The geese can look after themselves.

But I haven't been completely written off as father goose yet. When I run in hot pursuit after them and place myself at the head of the group of waddlers, Nemo is just fine with me taking over the lead.

14

Time to Go

THE PROBLEM WITH IMPRINTING ON PARENTS IS THAT NO matter how well or badly the attachment works, and whether you have a good or a bad relationship with your parents, there's no alternative. You have just this one set of parents, and somehow you have to deal with that—even when you're a goose.

Unfortunately, this is the case with Freddy and me. The goose is driving me nuts. Freddy either cannot or will not get used to me. He continues to refuse to listen to me, so he's always upsetting the dynamics of the group. After long deliberation, I've decided: it's Freddy or me.

THE TWO OF US ARE SITTING ALL ALONE ON THE DOCK AT the small swimming lake. I'm holding Freddy and looking out over the water. The air is warm but not oppressive. The atmosphere here at the lake is peaceful and almost festive. The late-afternoon sun is shining through the treetops on the other bank, making the water sparkle. Crickets are chirping, and a school of bream is hiding out in the shade under the dock. A duck lands on the lake, legs extended like a water skier.

"Freddy," I murmur.

I just can't bring myself to leave him in the wilderness. Yet that's exactly what I've decided to do. An hour ago I was certain that this is the only way. Freddy has to be released into the wild.

Freddy and me. We'd had problems from the start. I've thought long and hard about why that might be, but I haven't come up with any satisfactory answers. I've never heard of a rebellious phase in geese that lasts for months, and I'm positive that I didn't handle him any differently from the other geese, but whereas the others still more or less listen to me, after just a few weeks the effect of the horn on Freddy had almost completely worn off. And that's just one example.

The fact that Freddy's behavior breaks the group apart presents a problem. After all, the goal of our project from the beginning has been to use the geese as measuring devices that will follow me when I fly the ultralight. But as long as Freddy's here, I can't be certain that our flights will be successful. Maddin's company has done Freddy a lot of good recently, to be sure, and Maddin's gentleness has rubbed off on Freddy a bit, but even so, most of the disruptions in the group start with Freddy.

Yesterday, for example, I wanted to fly with the geese, and they all seemed really excited about the idea. Freddy started quite normally, but after about 150 feet, he veered off to the left in the direction of the aviary. The rest of the geese followed him, and I had to abort the flight. We just can't go on like this. If Freddy continues to rebel, it won't be long before I lose control of the other geese, as well.

For a few days, I even gave Freddy private flying lessons while the other geese were in the aviary, but his attitude didn't change one bit. It was just as ineffective when I did the opposite and flew with the other geese while he waited below. As soon as I shut him up in the crate at the airfield, he would honk with all his might, completely distracting the other geese. They would promptly break formation and land close to him.

It's all particularly disappointing because Freddy is an especially good and athletic flier. In contrast to Nemo, who takes a few wing beats to get up to speed, Freddy can lift off from standing like a Harrier jet. Freddy flies with magnificent ease and power, and the other geese could definitely learn a thing or two by watching him.

I remind myself that most families have their black sheep, or in this case, black goose. There's not much point in parents blaming themselves. But, of course, that's easier said than done. Was I somehow, consciously or unconsciously, less loving with Freddy after all? Perhaps my first reaction to his behavior was too extreme and intensified the whole conflict? Or am I now projecting too many human emotions onto Freddy? Is it perhaps a bit feebleminded to talk of the "drama of the gifted goose," just because it's a bit stubborn and rubs you the wrong way?

And, of course, the subject of blame is a bit more emotionally fraught with adoptive parents. I keep asking myself whether Freddy would have behaved completely differently if he'd had a real greylag goose for a mother. In the wild, would Freddy have turned out to be as loving and obedient as Paul?

Seven Reasons Freddy Has Grown on Me despite His Behavior

1. I never know if he's just kidding around
2. The grin on his bill
3. He's certainly not stupid
4. He's the best flier in the bunch
5. I can still remember how he used to peep at me
6. A Che Guevara goose costume would suit him
7. Who likes brownnosers, anyway?

• • • • • • • • •

THERE'S SOMETHING ELSE WORTH NOTING HERE. Precisely because Freddy and I butt heads so often, I've spent a lot of time with him; therefore, I've grown quite fond of the little revolutionary. Troublemakers always cause the most angst, but I can't imagine our little goose family without him.

"Freddy," I say again. "Weeweeweewee!"

But the goose doesn't react. He just looks pensively out over the lake. I try to convince myself that Freddy has never wanted it any other way. After all, he's always been against everything. And yet, he hasn't had a choice. When he hatched out of his egg, all there was—was me.

I dangle my feet in the water and watch the darting movements of the school of bream. Can I really let Freddy fly free? Should I? What about foxes, or dogs, or hunters? Have I taught him enough that he'll be able to get along on his own? Will he join a family of wild greylag geese? When I think about it, that seems unlikely. But he just can't stay with us any longer.

I put Freddy down on the damp planks and let go of him.

"Freddy, I'm sorry," I say. "I really believe that this is the best thing for you."

He looks at me calmly for a long time, as though he wants to say: *Are you serious? You want to leave me sitting here without any dinner? That's cold, man, that's really cold. But that's pretty much what I expected from you.*

"It's summer now, and you'll find things to eat all over the place. And if you don't figure it out, I'll come and collect you in a few days! Promise. But fly away now and have a great life out there!"

Freddy's wearing a solar-powered GPS transmitter around his neck. Using this small device, I can go online and check his whereabouts anytime. The transmitter is my assurance that I can catch Freddy again quickly if something goes wrong.

Great. Your promises are so genuine. Super generous of you to say that I can come back if I want to.

Freddy stays sitting on the dock and makes no move to fly away. He just waddles about ten feet farther, leaves his calling card on the wooden planks, and settles down again. Perhaps I should have used that old trick that sometimes works with small children—insist they do the opposite of what you want them to do.

"No, Freddy, you're going to stay right here! Don't even think of flying away! You're far too small a goose to manage that."

But suddenly I hear a honking overhead. A flock of geese is flying over us. Freddy responds by gabbling and giving one loud honk. Then he spreads his wings, and he's up and away. He glides low over the surface of the water. When the wild geese disappear over the treetops, Freddy lands in the middle of the lake and honks loudly after them. As if by magic, the goose formation returns, flies directly over Freddy, makes a wide turn over the lake, and lands on the water about two hundred yards away from him.

Freddy swims to the wild geese, and I feel a pang deep inside my chest. He paddles, his feet hidden under the water, leaving an ever-widening V on the surface. I feel as though I'm watching my son shyly approach a group of cool kids and carefully ask: "Can I play, too?" About a hundred yards from the geese, Freddy stops and looks at me and then back at the geese. I withdraw so that my presence doesn't make the other geese nervous. Of course, I'd love to stay and watch to see whether Freddy manages to make friends with the wild geese, but I don't want to get in the way.

Guys, that funny-looking, uncool dude over there's got nothing to do with me! I've no idea who he is!

"All the best, little one!" I call to him as I disappear through the undergrowth into the meadow and to the parking lot.

Back in the van, though, I begin to have second thoughts. I grip the steering wheel and stay sitting there for a while before I drive off. Did I do the right thing? Will the wild geese really accept Freddy? Or is he too proud even to ask them to take him in? Might the others think he's a freak because he's grown up around people?

• • • • • • • • •

FOR FOUR DAYS, I MONITOR FREDDY'S LOCATION USING the tracking device. To save energy, the transmitter sends coordinates only three times a day—all are close to the swimming lake. So it seems that Freddy hasn't managed to join the wild geese and fly off with them. Perhaps he was too annoying, even for them.

As it turns out, however, Freddy isn't completely alone on the lake. He's found a special lady friend, whom we meet a few days later on one of our goose walks. It's the baroness—the revolutionary has forged an alliance with the nobility.

"Have you lost a goose, by chance?" she's curious to know as she looks at the other, still reasonably loyal, geese waddling along behind me.

"No. No, I haven't."

"Well, that's strange. There's been a goose swimming along behind me in the lake for a few days now. It's wearing a solar collar and is so pretty and cute."

"Cute?"

"You're sure it's not one of yours?"

"Maybe."

"The poor little thing looks a bit thin. And so lonely."

"It's Freddy," I acknowledge, "but we didn't lose him. We hoped he'd join a flock of wild geese. He doesn't really fit in with our group anymore."

"Well, I think, you better help the goose. It's not going to make it on its own. And it's so trusting!"

"Thank you very much for letting me know. I'll look after the goose," I tell her, not quite knowing what I'm going to do.

Poor Freddy. Of all the people he could have chosen to save him, he chose a baroness.

I immediately get the van ready, stick the horn under my arm, and drive to the swimming area at the lake. I park the van in the lot and walk the rest of the way through the wood. When I can make out the swimming area in the distance, the scene is so surreal that I have to blink and shake my head. Freddy is sitting on the bank, completely relaxed, watching the children play.

As soon as he hears my horn, he honks loudly with relief, gabbles contentedly, and waddles in my direction. Freddy really has lost a lot of weight. The baroness was right: the poor chap definitely is on the scrawny side. I suddenly feel so guilty that I have to force myself not to start speaking to him in baby talk. The swimmers are already finding my strange behavior highly entertaining: a middle-aged guy with a horn arrives at the lake and crawls on all fours toward a greylag goose while muttering strange incantations.

"Come, Freddy. We're going home!" I whisper to him, as though I'd actually lost him instead of abandoning him.

Freddy is so relieved to see me again that he lets me pick him up and carry him to the van as though he were Paul. With the prodigal son sitting there next to me in the passenger seat, I don't know if I should apologize to him or admonish him.

• • • • • • • • •

WHEN I SHUT FREDDY IN THE AVIARY, I FULLY EXPECT TO see his siblings welcome him back. But Nemo and Calimero immediately lower their heads at Freddy menacingly.

I thought we were rid of this pain in the ass a long time ago! How come the old dude's dragged him back in here? Does it make sense to you, Calimero?

Because the dude's a wuss, that's why! I just don't want that asshole Freddy anywhere near me.

Freddy isn't the least intimidated by the two geese, so they attack him and bite at his feathers.

"Stop it, the pair of you. That's enough, you hear," I scold the self-appointed gatekeepers. "Leave poor Freddy in peace! Can't you see how thin he is?"

I have to wade in between them before they'll release him and step back.

Freddy is clearly agitated, but he soon calms down when I put a bowlful of grain in front of him and stay by his side. He pecks at the food with great relish, while I try to makes sense of his siblings' behavior. It never occurred to me that Freddy would be thrown out of the group after his short absence, and this complicates the situation even more. Geese can be pretty rough with each other.

Freddy eats until he's stuffed. I stay sitting next to him for a while, watching to see how the others behave. At first, they simply ignore Freddy. But as soon as he moves toward the group, he's driven off by Calimero, Nemo, and even Nils, who all advance on him with their heads held low. At least it doesn't come to blows. When they threaten him, he steps back.

After more than two hours, it is, of course, Freddy's close friend Maddin who cautiously reestablishes contact. As if he just happens to find himself near Freddy, Maddin begins to cautiously graze closer and closer to his best friend of old. Then he suddenly starts to preen Freddy's feathers, and the ice is broken immediately.

A preening frenzy ensues, and within a few minutes, Freddy and Maddin are thick as thieves once again. The rest of the geese watch the two turtledoves suspiciously, until finally it's Glorio who goes up to Freddy to welcome him back into the group.

We're just never going to get rid of Freddy.

15

Flying in Circles

ONE DAY IN SEPTEMBER, I NOTICE AN UNUSUAL NUMBER of feathers in a corner of the aviary. They're covert, or covering, feathers, and the geese usually don't shed many of them. This puzzles me, and I wonder whether the geese had a go at each other in the night, which seems unlikely. I think no more of it and set off to the airfield with them, as I do every day as long as the weather is reasonably good for flying.

But the next day, there are more covert feathers in the corner, and on the day after that, I'm pretty sure I could gather enough to stuff a good-sized pillow. I examine each goose's feathers carefully, and I can't find a good explanation for the condition of their plumage. But there's no denying it: the geese have begun to molt their covert feathers.

Molting is a natural process of losing and replacing feathers. Greylag geese molt once a year, because sooner or later their feathers wear out. Then they discard all their flight feathers as quickly as possible, and they all grow back in about four to six weeks. During this time, the geese are unable to fly and are particularly vulnerable. Therefore, they mostly keep close to the bank and take refuge on the water at the slightest disturbance, because this is the only place where most of their enemies can't reach them.

However, molting in September is fairly unusual for geese. Normally, the greylag geese in this area lose their feathers between the middle of May and the middle of June, right at the time when they're raising their goslings and don't need to fly much anyway. As soon as the goslings are capable of flying, the parents are over their molt, and they can give their offspring flying lessons using their brand-new feathers.

The majority of goose down used commercially comes from slaughtered geese. A small amount is down that geese lose in the natural, painless process of molting. There are, however, still countries in which live, conscious geese are plucked four to seven times before they are slaughtered. This way, goose down can be harvested much more quickly—and down ripped from live birds continues to find its way to the German market.

When I imagine what this must be like, it turns my stomach. I, for one, will never again risk buying down ripped from live geese. If I'm in any doubt, I'd rather resort to synthetic materials. And these fabrics offer many advantages over classic goose down. Modern fabrics are much better at insulating and wicking off water than natural down—to say nothing of them being nonallergenic.

Molting covert feathers—which cover and protect the flight and tail feathers—doesn't make the geese unable to fly, but it's an odd thing to be happening at this time of year. I talk to many goose specialists, and their unanimous opinion is that the molt could be contributing to the birds' unwillingness to fly right now. Molting is an energy-intensive process, and being energy efficient is one of the guiding principles in a goose's life—so why are they doing it? Perhaps the geese in my four-feather hotel are so well fed that they are molting simply because, in terms of their energy budget, they can afford to do so.

I gather up the feathers and worry. The missing covert feathers are not our only problem. On one of our training flights, Maddin is flapping around exactly where I want to land, and I end up brushing against him with the back wheel. The sound is like a fist punching a down pillow. Maddin is catapulted up into the air and crashes down onto the grass. At first, I'm convinced he can't have survived the blow.

I bring the plane to a stop and run back to the goose, who's lying on his back with his legs straight up in the air. But when I touch him, he gabbles just the way he usually does and even gets up. Maddin limps a bit, but other than that he seems fine. I decide to take him to the vet, just to be on the safe side.

In the waiting room, I'm reminded of the time when my son fell seriously ill when he was three months old. He was too young to tell us where it hurt or what was wrong, so we had a stomach-churning feeling of helplessness. I'm amazed that sitting here in the waiting room with Maddin on my lap, I'm reliving the fears I had at that time. Love and worry, it seems, transcend the species boundary. Love is simply love.

Maddin looks around curiously and isn't fazed by the Saint Bernard called Beethoven with its left leg wrapped in a bandage, or by Gonzalez, a fat gray tomcat sitting in a cage on its owner's lap, meowing most unhappily. He doesn't even become anxious when a little cocker spaniel called Flower, who just seems to have a bit of a cough, comes toward us wagging her tail. Maddin just keeps nibbling at the ties on my jacket. He finds the plastic cap that prevents the end of the cord from fraying particularly interesting. It doesn't take him long to chew it off completely. A goose's bill is incredibly strong.

Luckily, the vet can't find anything wrong other than a bruised bone. She prescribes a painkiller that I'm to squirt into his bill twice a day.

• • • • • • • • •

THE NEXT DAY, NEMO IS SITTING IN FRONT OF HIS BELOVED kiddie pool, barely moving. He appears slightly disoriented and doesn't greet me. When I pick him up, his left leg hangs limply. He seems to have lost all control over it. My first thought is paralysis as a result of a blocked or pinched nerve.

I feel Nemo's leg and it turns out there's something wrong with his thigh. The bony structures in his leg feel wobbly. I rush off to the vet again. Because it's still early in the morning, we're the first and, so far, only patients. The vet is quick to diagnose a broken leg. That means I have to take him for an X-ray. A joint veterinary practice a couple of villages over has a suitable machine. Nemo gets a mild sedative to make sure he doesn't fidget during the procedure.

As Nemo lies there completely limp, I look at the X-ray images. Nemo has sustained a complicated compound fracture to his thigh. In the wild, a diagnosis like this would be a death sentence, and we, too, must weigh what's to be done. The vets all agree that an operation is possible and worthwhile. Using a pin and a plate, the bone can be stabilized sufficiently to allow the bone to knit back together again.

I look from Nemo's motionless body to the vets, and back again, and consider my options. An operation on a goose. That's unusual and a bit bizarre, isn't it? We don't place much value on the life of a goose when we turn it into the centerpiece of our Christmas dinner. And think of all the people in the world who wouldn't be able to afford an operation like this, even if they had access to it. So what should we do? We can't just leave Nemo to his own devices. As father goose, I looked for and found him—or perhaps I should say, I found the egg that contained him—and that makes him my responsibility.

We're very lucky that the break is contained and Nemo doesn't have an open wound. This reduces the risk of infection. I decide in favor of the operation, and the very next morning Nemo goes under anesthetic. I learn later that it's not easy keeping a patient at a reasonably safe temperature during the procedure when he's basically lying on the operating table under a down comforter. Every once in a while, the vets have to step in and soak Nemo's feet in ice water so that he doesn't overheat.

Luckily, other than that, the operation goes smoothly, and our wildlife technicians build the recuperating goose a comfortable nest out of hay and straw for when he gets home. Here he can rest for a few days until his thighbone knits together again. The poor little chap.

Now you might think that the rest of the geese, his siblings, would be only too anxious to look after sick, wounded Nemo. But that would be a naïve fantasy. Even though the geese have a strong bond, they're all business when it comes to matters of hierarchy.

I have to put Nemo's comfortable convalescent nest in one of our travel crates; otherwise, the rest of the geese would probably attack him. As the highest-ranking but temporarily immobile gander, he's an easy target and can't defend himself. Calimero would certainly lose no time casting him off his throne, probably breaking even more of his bones while he was at it.

As Freddy disturbs our flight trials anyway, I leave him with Nemo and Maddin during the day so that the geese can keep each other company. The difference in rank between the three of them is clear enough that I'm sure Freddy and Maddin won't try to attack Nemo.

While Maddin and Nemo are recuperating with Freddy, I keep to the daily schedule of flight training for the rest of the geese. Often it's only thanks to this routine that I can keep myself somewhat motivated, but increasingly I catch myself thinking that perhaps I should simply call the whole project off.

The geese are still not flying high enough. They're accustomed to the ultralight and don't have any problem taking off with me, but I always end up aborting the flight. After two or three attempts to get them to fly higher, I usually give up and we drive home. I never thought I'd say this, but it's just no fun flying with the geese at the moment.

• • • • • • • • •

WHEN I'M ASSAILED BY DOUBT, I VISIT A GNARLED OLD oak tree in the wood. The circumference of the trunk is about twenty-five feet, and from below, the canopy looks gigantic. I lean against the trunk, shut my eyes, and imagine how much this tree must have lived through over the past three hundred or four hundred years. I think of how many dreadful or wonderful things must have happened while it was just standing here. The bark feels rough on my back, and I almost believe I can feel the gentle life force that has been flowing through the tree for so long.

When I was a child, I spent most of my spare time in the woods. My favorite game was what my father and I called "The Battle of the Pinecones." We pelted each other with large pinecones, and the person who scored the most direct hits won. In those days, I basically lived outside, and I played with whatever I could find on the woodland floor. That was enough for me.

Most times some of the peacefulness that radiates from the old oak settles into me, but the tree can't solve the problems I'm having flying with the geese.

Things are particularly bad on a big and important day for me. The film team is here again and wants to film from a helicopter. It's going to be difficult, especially because I have a black eye caused not by the gatekeeper Calimero but by little Nils. He unexpectedly flapped down to me and wanted to nibble on my hair, which he usually does quite gently and carefully. But all of a sudden, he grabbed my eyelid and began pecking at it. The pain was bearable, but my eye quickly swelled almost completely shut.

Despite my swollen eye, I load the geese into the van and set out for the airfield with Laura, who has arrived as backup specially for this event. We can hear the helicopter from a long way off as the machine appears over the mountains. There's an enormous camera mounted at the front on one of the skids.

When the helicopter's about half a mile away, I'm supposed to fly toward it. This makes me uneasy, because the rotors on the enormous machine create quite a wind. As I'm taxiing down the runway, I have the impression that there's a giant meat grinder churning away up there in the sky, whispering to me: "How about a quick back and sides?" The tops of the trees are bending over in the helicopter's downdraft.

The geese have no problem taking off next to me, and they line themselves up behind the right wing tip. Glorio is first, followed by Calimero, Nils, and Paul. The helicopter keeps coming closer. When I'm almost right in front of it, it veers slowly off to the side to allow us room to fly by. The geese have such confidence in me that they stay with the plane, completely indifferent to the monster.

However, we still don't make it any higher than sixty-five feet. Even at this low altitude, the geese do whatever they can to use as little energy as possible. They fly as efficiently as they can, which includes flying in formation as they've been genetically programmed to do. But they do more than just fly one behind the other—they also use the movement of the air around the ultralight to drastically reduce the amount of energy they have to expend.

When the plane is airborne, air flows up and over the leading edge of the wing, creating an updraft. After their second flight, the geese recognized that they could take advantage of this, and if they stay in this current of air, they barely have to flap their wings. They can practically surf the bow wave created by the leading edge of the wing. Thanks to me and my ultralight, they have something that I, as a hang glider, could only dream of: a wave of air streaming along with them that they can ride.

When the geese are surfing this wave above me, they're using up some of the updraft energy I need to stay aloft, so I have to add power to maintain the same altitude. As I do so, I watch Glorio leading the formation just off my wing. He's gliding along with his neck outstretched and an alert look in his eye. In contrast to his body, which is being rhythmically pumped up and down in the air as he beats his wings, his head is perfectly still, because his long neck almost completely offsets his wing movements. Glorio keeps looking back at his siblings. I get the impression that he's just waiting for Calimero, who's riding his slipstream, to finally relieve him.

How about you get your rear end up front here in the wind? Haven't you worked out that it's time to spell me off?

I'm so sorry, but that grain from yesterday still isn't sitting so well with me. Why's the dude flying so fast anyway? Does he have something to prove to these film people?

And Glorio does indeed suddenly fall back. Calimero takes over flying at the front, but then he suddenly turns away and starts losing altitude. We're flying over an enormous cornfield when I notice out of the corner of my eye that Paul is also descending. I try to at least keep Glorio and Nils with me, but Nils puts all his trust in Paul. I can almost feel Glorio wrestling with himself while keeping a watchful eye on his two younger siblings. Then he, too, turns away and flies after them, as though he wants to make sure they're safe. Now only my trusty soldier Calimero is left flying along by my side, and the other geese have disappeared into the cornfield.

I drop Calimero off on the ground and fly a wide circuit on my own, without a single goose. I search the cornfield with my naked eye, but I don't spot anything. The feathers of the geese camouflage them almost perfectly against a backdrop of grass or a cut field.

Then I spot Glorio, at least, from afar. Or, I should say, I spot the half-mile-long traffic jam he's created. He's decided to land on a busy road, and that's exactly where he wants to stay. A few drivers have got out and are standing next to their cars; others are honking their horns. I circle low over Glorio and see that a couple of employees from the city public works department are attempting to reason with the gander. The arguments advanced by the two workers fail to convince Glorio to move, however, and he stays standing right where he is.

"Catch him! Catch the goose! It belongs to me!" I yell down from the plane.

But I'm not sure if the two men understand what I'm saying. They look up and stare at me as though I were a prehistoric pterosaur. With the help of three other drivers, and doing a great imitation of a wolf pack, they manage to drive Glorio into the field alongside the road, where he magnanimously allows himself to be caught.

I yell as loudly as I can, "Airpooort!" and one of the men waves. I see the two climb into their city works van and bump along a track in the field in the direction of the airfield. But then it looks as though they're going to take a shortcut through a meadow that leads nowhere.

"Nooo!" I scream.

But the two men in the van can't hear what I'm saying. Are they planning to let Glorio go in the field? I have no choice but to turn round and land in the field. I make a low passing flight to check that there aren't any ditches or unexpected holes, and then I lose no time landing near the two puzzled men and my goose. Before I offer an explanation, I check Glorio for damage, but he doesn't seem to have been hit by a car.

"Thanks," I say to the two men.

"'Appy t' help. Thee's startin' wi' duck from t' field?" one of them answers in broadest Allemannic, and I can't help but smile. I love Allemannic, our local dialect, at least as much as I love the geese.

"'Tis no duck, 'tis a goose!" I explain. "Would thee be so kind as t' drive 'im quickly over t' airfield?"

The men look at the goose and say happily: "Aye, sure we would. Careful o' bumps when thee takes off from t' field, mind."

"Aye, no problem!" I reply.

Unfortunately, that's not completely true. Landing in a bumpy field is easy and even advantageous because the uneven surface helps with braking. Taking off is a completely different story. The more the surface slows you down, the longer the runway has to be. On asphalt, sixty-five feet is sometimes enough; on this field, however, I'm going to need more than three hundred.

I bid farewell to the two men and walk up and down the field a few times. It will do very nicely as a runway, and I'm happy there aren't any tall trees at the end of it. It's just a short hop before I need to land again and finally take delivery of Glorio.

• • • • • • • • •

NOW IT'S JUST NILS AND PAUL WHO ARE MISSING. THE TWO are fitted with tracking devices, but the signal's very weak in the middle of the cornfield, and that makes it difficult to pinpoint their location. The corn is already taller than I am, and it towers over Laura. As I part a couple of stalks and walk down the rows, after just a few steps, I'm pretty sure I've lost my field assistant, as well.

Deep in the middle of the green stalks, it's remarkably quiet. I call for Laura, Nils, and Paul, but none of them answers. Soon I can no longer tell how far into the cornfield I am, and I lose all sense of direction. Even if I jump up and down, I can't see over the tops of the plants. I feel like I'm in an endless ocean of corn.

I'm soaked in sweat and covered in mosquitoes. For a moment, I think that the tire tracks from a tractor will help me get oriented, but they just lead farther into the field. The soil under my feet is dry and dusty. I call Laura's name again, take two turns to the right, and sit down on the ground. At knee height, I get a better view, because the lower leaves have already fallen off the stalks of corn. And, indeed, four rows over, I spy a goose. It's Paul. I go and get him.

"I've got Paul," I call.

But Laura's still not answering. I stroke Paul's feathers and close my eyes. How difficult can it be to find your way out of a cornfield? Then the plants behind me suddenly open up like curtains, and Laura's standing beside me. I thought I was lost, but really I was barely fifty yards from the open meadow. We sit down on the grass and consider what to do next.

"What have you done with Nils?" I ask Paul, who's gently tugging on my hair. I can't imagine that Paul was in that cornfield alone.

We're thinking of getting going when we hear a honk close by. I don't believe it. No more than fifty yards away, the cornfield spits out Nils. Nils, the small still-too-light Nils, has found his way out of the cornfield all on his own.

Later we see from the location data on the GPS that Glorio flew two full circuits over the place where Paul and Nils landed before he veered off in the direction of the road where he caused the traffic jam. As the oldest, Glorio was worried about the other two and wanted to check they were okay.

16

Farewells

NEMO'S DYING.

I find him, one morning in October, lying in the aviary practically motionless. The upper part of his neck looks twisted, and he's wheezing alarmingly. It sounds as though he's trying to jabber quietly and with great effort out of a moist wound. I lift his body off the ground, talk to him reassuringly, and drive him to the vet. He hardly reacts. All he does is look at me with his dark eyes as though he's completely exhausted.

Of all of them, it had to be Nemo. I have no idea what could have happened to him. It looks as though he's broken something, but there's no reason for me to think that. Nothing points in that direction. Everything was fine last night, and his broken leg healed well some weeks ago. In the last few days he's been flying with us again, and he hasn't been limping at all.

The vet lays Nemo on a small white table in front of her, lifts his neck, and looks down his bill. In these surroundings, the goose looks so helpless. There's no sign of the elegance with which Nemo glides through the air. His feathers aren't moving and his feet are lying limply side by side.

The vet shakes her head and I understand. She can't do anything more for him.

"The little one has broken his esophagus," she says. "It's inoperable."

"But how could that have happened?"

"Difficult to say. He might have flown against the wire netting in the aviary or taken an awkward fall."

I grip the small table, force myself to take a deep breath, and don't know what to say to the vet. Is Nemo's death my fault? Should I have released him sooner, even though the project isn't over yet? Why would he suddenly start flying against the wire netting? Was Nemo perhaps unhappy being with me? In the wild, it's completely normal for at least one goose to die as the family grows up, but that doesn't make this any easier.

I bid farewell to the small lifeless bird by putting my hand carefully under his feathers to touch his body. Nemo is still warm.

"Take care, you old rascal," I whisper as I stroke his head one last time.

Seven Things We Don't Think of When We Eat Roast Goose

1. Their appetite for life
2. Their beautiful plumage
3. The feathers that sometimes stick out on their heads, making them look like kittens
4. Every goose means something to another goose
5. Their little goose personalities
6. The heartrending noises geese can make
7. How much goose shit there was in the gut of the bird we stuffed with plums

• • • • • • • • •

BECAUSE THINGS HAVE BEEN GOING SO BADLY WITH FLYing lately, I'm steeling myself to release the birds soon. I can't rush flight training, and they're increasingly anxious to leave me. It seems that a long flight way up in the sky with the geese will be nothing more than a dream for me now. If the geese don't want to do something, they won't.

I try to prepare the birds as well as I can for their future lives in the wild. Above all, they need to know how to get food without me and the catering services of the four-feather hotel. And so I take them for lots of walks in a nearby cornfield where the corn has recently been mown down. There are still some leftover cobs lying around in the field.

Although the geese enjoy corn, at first they don't know how to get to the cobs inside their leafy green packaging. So I sit down in front of them to show them what's hidden inside the leaves. I do this a few times. By the next day, the geese know how to peel the cobs using their strong bills so that they can peck out the kernels.

• • • • • • • • •

I'M STILL DEALING WITH NEMO'S DEATH WHEN MORE BAD news arrives. Heinrich, our caretaker, comes running up as I'm standing outside the Institute one afternoon.

"The dog," he gasps at me. "Micha, the dog!"

"What dog?"

"That dog that goes for walks along the track in the field! You must have seen it lots of times. It barked at the geese again."

"Jürgen?"

"What do you mean Jürgen? Who's Jürgen? No, I mean the dog. He ran up to the aviaries off his leash, and he's spooked the geese."

"Yes," I explain. "The dog's called Jürgen."

Because I still don't quite understand what's happened, I go down to the aviaries with Heinrich. I see what's up right away. Jürgen and his owner are long gone, but the dog must have scared Glorio so much that he muscled his way out under the aviary netting and flew away.

"I just caught a glimpse of him flying off in the direction of the pond," Heinrich informs me.

I set off for the airfield right away.

I taxi the Atos down the runway and fly directly to the three small ponds not far from the Institute. And, indeed, I think I can make out Jürgen and his owner on a path by the edge of the wood opposite. From a distance, the tiny dog looks sweet and harmless. I circle down over the ponds. In one spot, there are about a dozen greylag geese sitting on the bank. I fly multiple circuits over the geese, but I can't decide if Glorio is among them. I'm too far away and flying too fast.

And I can't find any geese in the surrounding fields or along the edge of the wood. On my way back, I fly over the three ponds again and watch as the little flock of greylag geese takes off and flies away in the direction of Lake Constance. I honk my horn and call into the air.

"Glorio! Is that you? Glorio! I'm so sorry about Jürgen!"

I even manage to catch up to the flying geese in my ultralight.

"Glorio!" I call. "Glorio!"

When I'm about sixty-five feet away, magically, one goose detaches itself from the group and flies over to me. It's Glorio. He stays close to me for about ten seconds, looks at me, and then rejoins the wild flock.

"Live well, Glorio! Enjoy your freedom!" I call as loudly as I can.

I watch him go, honk a couple of times to say goodbye, and turn sharply away.

17

The Passenger

TO CREATE ENERGY OUT OF NOTHING WOULD BE A WONderful thing, but sometimes you achieve more by being lazy than by weeks of effort.

It's already the middle of October and summer is basically over. Soon it will be too cold to fly, and that means our project will probably be a complete failure. Moreover, increasing morning fog, which is not unusual for Lake Constance, is complicating things for us. Periods of fog are the price we pay for the otherwise above-average weather here. But for someone who loves to fly, fog is the worst, for there's no escaping that soup. We often have to wait a long time until visibility is good enough to fly. The fog is sometimes so thick on the ground that we can't even do short circuits at low altitude. What I find particularly annoying is the thought that the sun is shining brightly above the fog layer, and the conditions up there are perfect for flying. But we simply can't break through the fog to get there.

I'm at the airfield again, sitting in my ultralight, and I don't know what to do. The air today is so damp that the Atos was clammy this morning and covered with a fine layer of moisture. Moisture on the leading edge can disrupt the wing profile. The effect is so noticeable that it feels to the pilot as though the wing won't work the way it's designed to. I used an old trick from my hang-gliding days and wiped the leading edge down with a cloth soaked with a rinse agent. The agent interferes with the surface tension of water so it can't form droplets.

For weeks it's been the same exasperating routine every morning. We manage two short circuits, but the geese don't want to fly any higher. Instead, they take their leave and fly back down to the field.

I ask you, why should we go up there when the best grass is down here?

Watch out. Soon he'll be telling us that the grass is even better up in the clouds.

I'm just not going to listen to him anymore.

The geese have finally grasped the fact that the flights serve my measurements but don't give them anything in return. To keep them guessing, I should have been taking off from a different airfield every time and landing at yet another one—perhaps then the geese would still believe that they have to cover some ground if they're to be rewarded with a juicy fresh field of grass.

Logistically, though, that just wouldn't have been possible. Unfortunately, in Germany, you can't just land an ultralight in any field that takes your fancy unless it's an emergency, and the next airfield is six miles away, which is too far for the geese at their current level of training. Normally, geese can fly such a distance with a shake of their tail, but not my all-inclusive–resort geese.

And so we always take off from the same airfield and land back there again. For a long time now, the geese have probably thought of this circuit as a meaningless exercise in marching on the spot. Even my idea of motivating the birds with food by letting them go hungry overnight in the aviary and then feeding them after the flight at the airfield hasn't worked.

• • • • • • • • •

THE GEESE ARE GRAZING IN FRONT OF THE HANGAR. ONLY Paul is still standing loyally next to the airplane—or perhaps he just doesn't know what else to do. I really should call it a day. But I don't want to get out of the cockpit and tie the plane down for the night just yet. I wonder if I should try it with just Paul. After all, he's the one who's always trusted me the most. Perhaps he might be willing to go it alone and fly behind the Atos, even if the other geese aren't here. It's not terribly likely, but I don't have a better plan at this point. The most important thing is that we don't keep repeating the same dispiriting trial flights.

If I'm going to do this, I'll have to taxi the Atos into its starting position and then hope that Paul waddles along behind me so that he can start next to the wing. But that seems too tedious and way too much work, so I simply put Paul on my lap in the cockpit and we taxi to the start position together. He immediately begins gently nibbling on my nose with his bill. If it were Freddy, I'd have to wait a good long time for an expression of affection like that.

"Oh, Paul," I whisper sadly in his ear, "I'm all out of ideas. You geese simply don't want to participate any longer, and I can't force you to. What do you think? Would you, at least, like to fly a proper circuit with me?"

I use my right hand to steer and accelerate while holding Paul tightly on my lap with my left. We taxi slowly and deliberately to the head of runway 01.

Calimero, look back there at what the dude and the suck-up are doing!

They can fly around the world for all I care.

I actually don't need to hold onto Paul very tightly, because he's so relaxed that he stays sitting calmly on my lap. He makes absolutely no effort to break loose from my grip, and he doesn't look in the least bit concerned. Suddenly, an idea begins to take shape in my mind. Taking off with only one hand on the controls doesn't exactly check the boxes as a safe launch, but what the heck. Let's go for it.

The plane lurches forward, we rumble over the field, and after just a few yards, we're up in the air. Just the three of us: the Atos, Micha, and Paul. Paul doesn't flinch, and he doesn't flap around nervously, either. He just looks around inquisitively—he seems to really enjoy being in the cockpit.

I even let go of him so that he's sitting on my lap completely unrestrained. He calmly nibbles my beard and the knotted nylon cable attached to the wing flaps, or simply holds his streamlined head into the air rushing toward us. We climb higher and higher, until suddenly the whole panorama of the Alps and Lake Constance opens up in front of us.

Seven Reasons Why Flying Is Addictive

1. The unreal quiet when you take off
2. Adrenaline
3. Riding suddenly becomes soaring
4. The little darling basin of water that is Lake Constance
5. The tiny toy cars and toy trains
6. The poppy fields are at their most beautiful from this vantage point
7. Wondering what it is that makes everyone down there so stressed

• • • • • • • • •

WE REACH AN ALTITUDE OF ABOUT 4,600 FEET ABOVE SEA level. My idea is to simply throw the goose out of the plane. I'm traveling at forty miles an hour, which is in the ideal range for a goose, and with the head wind, Paul won't have any problem coping. Can a fish drown if you throw it in water? All Paul has to do is stretch out his wings and fly.

I slowly reduce speed, the propeller keeps turning behind us, and everything goes quiet. Up here there's only the sound of the head wind. Paul's sitting on my lap, just going along for the ride and looking up into the sky. My arms hang in the emptiness to either side of the pilot's seat.

I feel transported and disconnected from everything that's happening in the strange lives of the tiny people down below. As strange as it sounds, at this height, in the middle of nothing, I feel grounded. Nearer to myself. Should I really let the goose fall into the yawning void below?

"Shall we fly together?" I ask the greylag goose, looking deeply into his dark eyes.

The Atos glides along, as though it were riding a set of elevated railroad tracks. Paul suddenly stands up, takes a quick look down over the edge of the seat, and jumps into the air of his own accord. He falls for only a few feet before extending his wings, and he immediately begins flying alongside me with powerful strokes. He's so close that I could reach out and stroke his tail feathers as he flies.

For the first time, I have the time to observe the flight movements of a greylag goose with nothing to distract me—a perfect symbiosis of feathers, aerodynamics, and molecules of air. Now I can also finally check how the measuring device sits on the back of a flying goose and whether the little pitot tube we're using to measure airspeed is positioned correctly as air streams over it.

All the flights until now have been rather nerve wracking. This flight, in contrast, is like a high-altitude stroll. A colorful fall landscape is drifting by beneath us, and the sun, now low in the sky, is bathing everything in a warm golden glow. I'd like nothing more than to float like this forever. Paul's neck is stretched out in front of him yet rests peacefully in the air while his wings beat powerfully and elegantly. His dark eyes radiate pure contentment.

Mind you, the temperature at this altitude in October is only about forty degrees Fahrenheit. As glorious as this flight is, my hands are slowly turning numb, and I've lost almost all feeling in my fingers. The head wind, which is blasting in my face and through my rather too-thin flight gloves at nearly forty-five miles an hour, is a good part of the reason that I'm slowly but surely getting chilled to the bone. Paul is far better protected in his down coat. The wind doesn't seem to bother him at all. I think I even see him shoot me a questioning glance when I engage the flaps and begin a quick descent to the airfield.

The ultralight seems clumsy compared with the goose. I observe at close range the confidence with which Paul alters the shape of his wings to lose altitude and stay balanced at the same time. He forms an upside-down U with his wings while simultaneously spreading his feet wide and stretching them out in front of him, descending to the field just as quickly as I do. I have another opportunity to wonder at a familiar maneuver: while flying along, Paul turns onto his back at lightning speed. For a few seconds, while he's upside down, he plummets straight down to earth.

We land next to the van at almost the same time. I'm over the moon. From now on, we can take our measurements at whatever height we like. Who would have thought that my laziness would open up such fantastic possibilities for the project?

I'm so excited I've completely forgotten about the rest of the geese in the grass around the hangar. Even from a distance, I can see that there aren't enough of them. Maddin and Freddy are missing. Yet again, you could say. And again, I spend many hours honking and calling for the two geese. But without knowing why, I have a distinct feeling that the two of them won't turn up so easily this time. Freddy and Maddin are grown-up geese, and I hope that they can make it out in the wild without me.

At least after my success with Paul as a passenger, it's not so important to find Maddin and Freddy. As I always do, I keep driving around for a while in the van, honk for the two, and call, but they don't turn up. My gut tells me that this time they're gone for good, and I'm happy that Freddy has his good friend Maddin by his side.

• • • • • • • • •

THAT EVENING, AS I SIT AT MY LAPTOP AND PLUG IN THE data logger, my hopes are confirmed: the readout from the measuring device yields an enormous amount of data about acceleration, position, atmospheric pressure, direction, and speed. This is exactly the information we've been interested in from the outset—raw data on how a goose flies.

Collecting data on a logger might sound boring as a research goal, yet it's anything but. These are data that simply didn't exist before. To capture "inside information" and get it from way up in the air—that's always been just a dream. Until now.

Now that we know that this is possible, in future we will be able to use geese as living, airborne measuring devices. We will be able to fly to targeted layers of air or areas of turbulence and investigate these weather phenomena using birds. For example, no one yet knows how a goose alters its flight mechanics and wing geometry in response to heavy turbulence, or whether it exploits upward-moving currents of air when it's flying straight ahead or prefers to avoid them.

Of course, it's important that we don't gather these data just from Paul but from more geese if possible, because they all fly slightly differently. Apart from that, it would be good to have the data loggers measure not just one goose flying on its own but a number of geese flying in formation.

A few days later, I have an idea for how I can use a similar method to get the other two remaining geese up into the air: with a powered paraglider and two baskets that I'll mount to the right and left of the pilot's seat. After the many weeks of stagnation, during which I've barely managed to get the geese a hundred feet off the ground, suddenly everything is moving along very quickly. I talk this over with the film company, which still hasn't finished the film for ZDF because it's been so difficult to shoot at such low altitude. And, finally, I meet with Max, a rail-thin and completely laid-back paraglider pilot, who's willing to help me with filming and taking measurements.

Max's three-wheeled paraglider looks a bit like a tricycle capable of taking to the air. We put Calimero and Nils in two plastic baskets with holes in the sides to the right and left of the cockpit. The pair looks as though they've been loaded onto a terrestrial trike and are just waiting to be chauffeured around the corner to do some shopping. The geese find high-altitude travel in the baskets a bit of a challenge, but by the second test run, they relax and sit calmly as they allow Max to fly them around in the air.

Paul gets to sit on my lap again in the ultralight, so after two weeks of training, we develop the following protocol. When we reach the desired altitude, Max positions his paraglider about two hundred feet above the ultralight. Then I release Paul. Amazingly, the other two geese immediately try to join Paul and me. They fly down from their baskets and arrange themselves in a small formation.

• • • • • • • • •

WE CAPTURE IT ALL ON FILM ON THE LAST SUNNY DAY before winter really sets in. And so we fly—Calimero, Nils, Paul, and I—with a paraglider and a helicopter in tow, with good visibility and an expansive view over Lake Constance. The helicopter circles us, keeping a safe distance. I hardly notice it. I'm far too absorbed in the experience of us all flying together so high up in the sky.

Like a sponge, I'm absorbing the bright colors of the trees, the smell of the land, and the feeling of belonging with the geese. I know that our days together are numbered, and that we probably won't get to experience many more flights like this. Unfortunately, my fears are realized soon after landing.

While the geese are fortifying themselves with grain in the temporary aviary where I've gathered them, the sound of the helicopter propeller suddenly thuds behind us far too loudly. The fog has become denser, and visibility is very bad. How's he going to land? The pilot is flying by sight, which means that he's using the buildings to orient himself. He heads directly toward me and would have landed perfectly—if the aviary hadn't been directly in between us. The noise of the motors becomes a deafening roar. I wave and gesticulate, but it's too late. The downdraft from the helicopter blasts the geese right out of the aviary, and all I can see is the geese flying off in all directions.

"Are you completely mad?" I yell as I run toward Max, who has witnessed the whole thing.

Max rushes off to the end of the runway and disappears into the fog. I don't know which way I should run, so I stay where I am. Just a few minutes later, I hear a voice. It's Max.

"Over there, under the old cowshed, there's two geese!" he calls.

He sounds relieved, but I still can't calm down.

"Goddamn it!" I yell into the fog.

At that moment, I see a goose flying through the milky gray sky. Its wings seem to be beating slowly and effortlessly. It's Paul. I call and call, but Paul doesn't turn toward me, and in no time at all, he's disappeared into the fog. Under the overhang of the old cowshed sit Nils and Calimero, my last two geese.

We look for Paul for a few hours, but we come up empty handed. I blame myself that he left, because I insisted on filming with the helicopter.

Despite losing Paul, I fly up into the air once more with Nils on my lap so that we can finish filming. I release him into the air from the plane. He's so close to the cockpit that I can see the dark pupil in his eye. He flies just as weightlessly and contentedly as Paul. Using Nils, we manage to gather the last of our data and complete the film work. So at the very end, Nils Holgersson was the one who took off with me and flew up into the sky.

18

Back to Civilization

I HAD IMAGINED A VERY ROMANTIC FAREWELL FROM THE geese. I saw myself granting them their well-earned freedom in a celebratory ceremony at some idyllic spot by the lake after the successful completion of the project. In my somewhat weaker moments, I even saw the geese looking back one last time as they flew away, a twinkle in their eyes as they wink to thank me for taking care of them. But no matter how much I might have sometimes wished they were, my geese aren't cartoon characters but wild animals. They have minds of their own, and since they've been able to fly, I haven't been able to tell them what to do.

The very idea that I would first have to teach the geese how to fly was somewhat naïve. There's no question that the geese learned some things from me, but they knew how to do a lot of things all by themselves. And I hadn't anticipated that for some things it would be the other way around, that I would be the one learning from the geese.

Nature taught me an even better lesson than I could ever have taught the geese: nothing is predictable and everything is in a dynamic flux. Nature, I learned, is brutal and wonderful at the same time.

Of course, not all our setbacks can be laid at nature's door. Thanks to our flight operations and our research project, we also exposed geese to danger. Perhaps out in the wild, Nemo would not have died, but nature is not idyllic. If a mother goose has four eggs that hatch, sometimes not a single gosling survives, so numerous are the dangers and enemies geese might encounter in their first months of life. The quota of survivors, then, was somewhat higher with me than with a real mother goose.

I knew all of this beforehand, but I hadn't really taken it in. It's hard for a human father goose to accept that some of the geese that he's laboriously raised by hand with such affection won't make it to adulthood. As people, we're not used to that—for the geese, however, that's life.

• • • • • • • • •

BEFORE I RELINQUISH THE CAMPING TRAILER AND FINALLY end my goose adventure, my human children, Amélie and Ronin, come to visit one last time. They often visited me and the geese in the camping trailer and were very fond of my animal children. With Nils and Calimero—my two last remaining geese—we make a trip to a pond in a field nearby. In the big field next to the pond, there are chocolate bars for the children and grain for the two geese.

We lie on a large picnic blanket under the pear trees and look up into the sky. A single cirrus cloud is moving so slowly that it looks like it got stuck in the middle. We listen to the sounds of bills plucking blades of grass. I think back to many moments I've shared with the geese: how they would whistle sleepily up close to me; how their moods often mirrored mine; how Nemo, when he rose to be one of the leaders, would lie on my stomach as though it were his throne; and how Paul would slip up under my sweater when he was a little gosling. How beautiful and simple things were then. The air smells of hay, fruit, and goose poo, and all of a sudden I feel tears of pure joy welling up inside me.

Why is it so difficult for us to accept things just the way they are? And why isn't it difficult for the geese? Why have I butted up against facts that I simply cannot change? I sit up, look at Calimero, and know: I hadn't planned it this way, but in the past few months, the geese have released, or opened up, something inside me. Perhaps you could even call them therapy geese. Seven little greylag goslings have helped me find myself again after years of being lost and shown me what really matters in life: loving others and loving life itself.

With the geese, I could simply *be* for months on end. Without expectation or judgment. This created in me a sense of emotional freedom that I've never experienced before. I feel a lump in my throat, but then Calimero comes waddling over, nibbles gently on my hair, and poops, missing the blanket by just a few fractions of an inch.

• • • • • • • • •

"THEE TAKIN' THEY GEESE HOME WI' THEE?" ASKS THE renter of the camping trailer when I hand over the keys.

"No, no," I explain. "They're all grown up and can manage on their own. There are only two geese left anyway, and our experiment is over now."

"Thee had fun wi' they geese?"

"Yes, of course. But it was much more than that."

"Wha's that then?"

"It's complicated."

"Wha's complicated about a goose?"

"You'll just have to try it for yourself. Living with geese. Really. Everyone should do it at least once."

The renter looks at me a bit perplexed, shakes his head, and then climbs into his metallic-blue jeep to pull the camping trailer out of the field. I stand on the grass with Calimero and Nils, and wave.

Hey, Calimero. What's happening to our house?

Just look! Perhaps we should have listened to Freddy. He was always telling us there was something odd about this guy.

A few minutes later, all that remains of the camping trailer is a bald brown rectangle on the field. The tamped path laid down by Fridolin stands out particularly starkly now. I collapse the folding table and bench. The nest belonging to Ilse and Horst has been empty for some time now. And the cuckoo has disappeared, as well.

My nights in Duckingham Palace are over, and I must bid farewell to my Spartan life, my beer benches, and my deck. Saying farewell to all this also means that I can finally move back home, and I will be able to move freely once again, without being followed by geese wherever I go.

• • • • • • • • •

THE ONLY ONES IN THE AVIARY NOW ARE NILS AND Calimero. The rest of the geese have departed, all but one into the wild. I'm not sure all of them will make it out there. And that's why I want to do things right with Nils and Calimero. Unfortunately, I can't take the two of them home with me, because geese don't make good pets. They can never be housebroken, and that's just the way it is.

However, now that it's winter, I can't just abandon them. It would be too difficult for them to find food in harsh conditions. Someone who's spent his whole life in a four-feather hotel can't be expected to survive if he's suddenly kicked out onto the street in winter. Therefore, Calimero and Nils stay in the aviary a while longer, and I visit them both as often as I can. Then I take them over to a wildlife park nearby.

They won't live completely independently in the park, as they'll still be fed, but they'll be well taken care of.

They can fly away at any time if they want to, and if they don't want to leave, the park offers them proximity to people, which is what they've grown accustomed to. The park is a favorite spot for families, and along with goats and sheep, children can admire Nils and Calimero in its small flock of geese.

On a beautiful sunny spring day at the beginning of March, the time has come to introduce Nils and Calimero to their new home. I open the door to the travel crate in the big field at the wildlife park in Weiher to let out the last two geese. Calimero and Nils waddle out uncertainly and look around.

I stay sitting next to them for a while. I don't have to wait long. It's Calimero who marches fearlessly up to the strange geese as they check him out. Little Nils waddles after him, and neither goose looks back.

I watch them go and think that soon Nils and Calimero will start families of their own. I hope they'll teach their children everything geese need to know: that it's better to sleep in the middle of the lake in case a fox attacks in the night, that you can find delicious food in a cornfield, and that that dude over there with his odd-looking flying contraption isn't so bad after all.

● ● ● ● ● ● ● ● ●

MY FIRST, AND BY NOW COMPLETELY UNFAMILIAR, INTERaction with civilization is—shopping. I need provisions for my apartment.

I pull up in front of a discount grocery store and think how impractical it is that people need to shop for everything. They have to laboriously look for what they want, and then put the stuff in their shopping cart and transfer it to the trunk of their car. They have to carry it into their houses themselves, and finally, they still have to wonder what they're going to cook for dinner with the groceries they've bought.

My advice as father goose is simply this:

"People, just walk out of the house and into a field and nibble on the grass." Or "Ahem, enjoy the grass! It's juicy and delicious!"

Then I realize that I'm not a goose any longer. And even in the camping trailer I didn't eat grass. Instead, I relied on the assistance of our caretaker, Heinrich, and my friends, who brought me food. But now I'm a normal person again, so I have to go shopping. I need food. There's no way around that.

I put a euro in the little plastic slot and free a shopping cart. The cart has a pronounced list to the right, but that doesn't bother me. I enter through the automatic doors and smell the air conditioning. I can already feel anxiety spreading through me. It's so very hectic and stressful, and there are so many items on the shelves.

People are buying things like there's no tomorrow, and a couple of months ago that would probably have been me, as well. I could never have imagined how little I would miss any of this. People are buying eggs, chicken, and duck, completely unaware of how much life and diversity there is in what they're buying. Without asking themselves if the bird they're going to serve for dinner is stubborn or cuddly or particularly passionate about water.

I push my cart through the store, feeling as though I'm walking through a warehouse full of zombies. I can see their empty, stressed-out gaze. And then there's the latent aggression when you just happen to stop that one time in the middle of an aisle with your cart and look back at them. There's the video monitor from which an enthusiastic voice sings the praises of some probably quite useless product. People are throwing vegetables into their carts without having any idea what kind of world these products come from or how much work it takes to grow a single beautifully formed cauliflower.

Okay, I'll admit that maybe I'm exaggerating a bit, but I'm only halfway through my shopping when I'm overcome by an urgent need to escape from the store as quickly as possible. But even that act is stressful, because the automatic door only opens in the opposite direction and I can't just bypass the checkout with my cart. I've felt this way before—after the obligatory Saturday shopping expedition with my girlfriend, a marathon of purse holding and fashion advice. I was left with the feeling that when we acquire things, we lose far more than we gain: our equilibrium, our zest for life.

I put a few necessities in my cart and watch an older, fragile-looking woman. Holding tightly onto her cart with one hand, she's stretching her other hand up to reach a jar of jam on one of the upper shelves. Her bent body is straining under the effort, but she still can't reach the jam. She's upset for a while, tries again, and then gives up. Finally, out of necessity, she takes a more expensive product at eye level within easy reach.

"Excuse me, but did you want this jar?" I ask in a friendly manner, reaching up to the upper shelf.

I offer her the jam and smile.

"I'm old but not helpless," she snarls at me and adds for good measure, "Go to hell!"

I'm so flummoxed that I put the jam in my own cart and barely manage to mumble an apology. Staring straight ahead, I push my cart directly to the checkout, keep my eyes fixed on the conveyor belt, pay as quickly as I can, and try not to annoy anyone else. Why are people so complicated? I have to get out of here. I want to go home. Home to my geese.

19

The Birds Have Flown

FREDDY IS PROBABLY STILL DREAMING OF REVOLUTION and nurturing his contrary nature in a group of eight to ten geese that often hang out around a beach café in Radolfzell. With persistent jabbering and begging looks, they badger customers for bread, cakes, and other treats. Ironically, the goose who always wanted to be particularly independent is now dependent on the generosity of others—the maverick has become the beggar. Sometimes I blame myself about Freddy, and then I see him in front of me, how he accepts handouts with a look that suggests he finds them offensive, yet he snaps them up greedily and whispers to the other geese in his flock: *I know my way around people. I know all about them. Me and the little one over there, we've lived with a real person. He sat around on his bench the whole time and thought he was our pa. But I noticed right away that there was something off about him.*

MADDIN is more critical of Freddy's revolutionary talk these days, but he hasn't left him, and he still marches along behind him, loyal as ever. He, too, belongs to the group of begging geese at the beach café. *Hey, Freddy, I don't mean this personally, but if everything's so unfair in the wild and that's why we're here begging, how come you always get the biggest share?*

Shut your bill, Maddin.

Sure, Okay. Sorry. I was just sayin'.

CALIMERO now calls the Lochmühle leisure park in Eigeltingen home, and you can visit him there. There aren't any enemies for Calimero to fight here, but when children and adults call him "sweet" or "cute" and want to stroke "that adorable goose," perhaps it helps him to remember his wild days in the camping trailer and his Rambo times with me.

NILS, the youngest and smallest of the geese, also ended up at the Lochmühle leisure park, where he fit right in with the motley crew that make up the flock of geese there. Delicate little Nils feels safe in this group of geese. His big brother Calimero is usually around and always ready to protect little Nils by lowering his head, stretching out his neck, and hissing.

PAUL, my favorite cuddliest, gentlest poster boy of a goose has vanished into the wild. I've never seen him again, and I've never had any news of where he might be. I once thought that Paul would do anything I wanted him to do, but he hasn't yet come to visit me. Perhaps Paul, who makes new connections easily with his friendly demeanor, is far, far away by now, with his wild goose friends. I just hope that his openness has not been exploited by a fox or another goose on an ego trip.

Nobody gave a eulogy for NEMO, but I've sort of started one: Nemo loved water, food, and my teeth. He was a big goose who had his ups and downs, and few geese could resist his charismatic personality. He soon became a leader, even as a gosling, but his weakness for dandelions was his downfall, and he had difficulty getting off the ground. But Nemo was up to the challenge. He fought his way through with his singular determination and was soon flying along right up front. His death came suddenly and far too soon.

Twice, I saw GLORIO, the eldest of the goose brothers, on a small pond next to the airfield. He took off in the midst of a group of wild greylag geese, and he didn't stand out in any way. His wings carried him up into the air as fast as those of his friends and colleagues. He looked like an absolutely normal wild-raised goose, and it's only thanks to the pink leg band clearly visible from a distance that I could tell it was him. I don't know if he noticed me, but I lifted my head, thought of the little baby goose under my sweater, and watched the geese as they flew away until they were nothing more than black dots on the horizon. For one brief moment, I was proud just like a real dad.

THE DATA we collected with the geese was amazing. In more than sixty flights, we gathered many megabytes of information—enough to be the basis for multiple PhD theses. Thanks to the data loggers carried by the geese, we now have information on wind speed and direction as birds flap their wings in flight, and data on how they fly. Now that we've shown that data loggers can be carried without interfering with bird flight, researchers will be able to use them to collect meteorological data from hard-to-reach places in the future , and they will also be used as part of a larger project being undertaken by the Max Planck Institute to track bird migrations around the world.

Afterword

MICHAEL QUETTING'S PROJECT WAS PART OF A MUCH LARger endeavor, called ICARUS (International Cooperation for Animal Research Using Space), being undertaken by the Max Planck Society, the Russian space agency, the German Aerospace Center, and the University of Konstanz.

In February 2018, a Russian *Progress* supply ship delivered the ICARUS receiving and transmitting antennas to the International Space Station. If all goes according to plan, by August, two Russian cosmonauts will have undertaken a five-hour spacewalk to attach them to the outside of the station, allowing ICARUS's work to begin in earnest.

Using tiny data transmitters, ICARUS will analyze animal movements around the Earth to track bird migration, the global spread of pathogens, the efficacy of pest-control measures, population levels of endangered species, and habitat shifts in response to climate. We may even be able to definitively identify animals that can sense impending natural disasters, and thus be able to save thousands of human lives.

Michael and his geese were involved in an important initial testing phase. Calimero, Gloria, Nemo, Maddin, Frieda, Paula, and Nils were enchanting individuals, but they were also vital contributors to what will be a far-reaching and life-changing project for animals, humans, and the planet we all share.

DR. MARTIN WIKELSKI, Director, Max Planck Institute for Ornithology, Radolfzell, and head of the ICARUS project.

Acknowledgments

MARTIN WIKELSKI, THE DIRECTOR OF THE MAX PLANCK Institute of Ornithology. You made this project possible in the first place, thanks to your ideas and vision, your idealism, your and wealth of scientific knowledge.

My children, Amélie and Ronin, who helped me enthusiastically every second weekend and happily shared their "dad time" with the geese.

My parents, Bea and Edwin, for taxiing the children to and fro, buying groceries, and doing laundry. Your moral support has always motivated me to never give up.

All my friends and acquaintances who stood by me at this extraordinary time and supported me with visits and caring words. Above all, Eva.

Toni Roth, the builder and developer of the sensational carbon tandem trikes. Without your Gyro Gearloose ideas, I would have failed miserably!

All my coworkers at the MPI for Ornithology in Radolfzell, especially those who work in animal care. You do a fantastic job every day.